High-performance CCD cameras have opened up an exciting new window on the Universe for amateur astronomers. This book provides a complete, self-contained guide to choosing and using CCD cameras.

The book starts with a no-nonsense introduction to how a CCD camera works and just what determines its performance. The authors then show how to use a CCD camera and accurately calibrate the images obtained. A clear review is provided of the software available for visualizing, analyzing, and processing digital images. Finally, the reader is guided through a series of key areas in astronomy where one can make best use of CCD cameras.

This handy volume is packed with practical tips. It provides a clear introduction to CCD astronomy for novices and an indispensable reference for more experienced amateur astronomers.

D1262187

The Practical Astronomy Handbooks are a new concept in publishing for amateur and leisure astronomy. These books are for active amateurs who want to get the very best out of their telescopes and who want to make productive observations and new discoveries. The emphasis is strongly practical: what equipment is needed, how to use it, what to observe, and how to record observations in a way that will be useful to others. Each title in the series is devoted either to the techniques used for a particular class of object, for example observing the Moon or variable stars, or to the application of a technique, for example the use of a new detector, to amateur astronomy in general. The series builds into an indispensable library of practical information for all active observers.

Titles available in this series

A Practical Guide to CCD Astronomy

A Practical Guide
to CCD Astronomy

PATRICK MARTINEZ AND ALAIN KLOTZ

translated by André Demers

PUBLISHED BY THE PRESS SYNDICATE OF THE UNIVERSITY OF CAMBRIDGE
The Pitt Building, Trumpington Street, Cambridge CB2 1RP, United Kingdom

CAMBRIDGE UNIVERSITY PRESS
The Edinburgh Building, Cambridge, CB2 2RU, United Kingdom
40 West 20th Street, New York, NY 10011-4211, USA
10 Stamford Road, Oakleigh, Melbourne 3166, Australia

© Cambridge University Press 1998

First published 1998

Printed in the United Kingdom at the University Press, Cambridge

Typeset in 9/12pt Meridien [SE]

A catalogue record for this book is available from the British Library

Library of Congress Cataloguing in Publication data

Martinez, Patrick.
 [Guide pratique de l'astronomie CCD. English]
 A practical guide to CCD astronomy / Patrick Martinez and Alain
Klotz: translated by André Demers.
   p.   cm. – (Practical astronomy handbook series; 8)
 Includes bibliographical references.
 ISBN 0 521 59063 9 (hardbound).
 ISBN 0 521 59950 4 (pbk.)
 1. Charge coupled devices – Handbooks, manuals, etc. 2. CCD
cameras – Handbooks, manuals, etc. 3. Imaging systems in astronomy –
Handbooks, manuals, etc. I. Klotz, Alain, 1967–. II. Title.
III. Series.
QB127.4.M37   1997
522'.63–dc21 96-54304 CIP

Contents

Foreword

Sky aficionados, whether professional astronomers or amateurs, always have two preoccupations. One is aesthetic: they want to capture, in a memorable way, the beauty of the sky. The other is, of course, scientific: they would love to quantify their observations, compare them with others', and verify or discover new effects.

Everyone knows that an astronomical observation uses a complex system. The telescope is an essential element, but is not unique: there is also the choice of site, shielded from light interference and turbulence, the construction and thermal stabilization of the dome, and, of course, the light detector that controls the quality of the final image. It would be more appropriate to speak of the 'observing system', whose every link is essential.

In its time, J. Texereau and G. de Vaucouleurs' famous book *L'Astrophotographie d'amateur* inspired generations of amateurs when photography was the best way to capture photons. Today, modern light detectors are charge-coupled devices, commonly known as CCDs. If their format does not reach that of a photographic plate, still unequaled in the number of pixels it offers, their sensitivity is several dozen times better. And since we all know that the time needed to reach a given signal-to-noise ratio (which is directly linked to the possibility of detecting a possible astronomical source) varies as the inverse square of the sensitivity, it is easy to understand the incredible leap forward CCDs will enable observers to make. Today, observers know that progress in CCD technology – micro-electronics, or even nanoelectronics – is still being made: read noise is decreasing, as are costs, the number of pixels is increasing, and soon it will be possible to tile the focal plane of the large 8 to 10 meter telescopes, now being constructed, with CCD mosaics, leading to hundred million sensitive points, which begins to approach the performance of a photographic plate.

It would be unfortunate if amateurs remained uninvolved with this trend, since not only do CCDs provide these images at reduced exposure times, or deeper magnitudes, but images are directly available in a digital form. Coupled with a personal computer, the CCD opens a whole field of image treatment, up until now reserved for professionals: filtering, formatting, quantitative measurements, visualization in real or false colors, stacking, archiving, and establishing a database.

Without forgetting the extraordinary possibility the digital image offers, through its televised projection, it can be displayed right away or kept for later viewing. No more lines of curious people passing by the eyepiece, too inexperienced to twist their neck or use the right eye, or knowing how to appreciate the darkness and savor a trembling image in the eyepiece. Disappointments expressed as an 'I can't see anything!' followed by an upset observer enthusiastically trying to explain what he 'can' see are almost gone. A live sky on a big screen is another achievement of the CCD that is continuing to have an impact in the clubs and on school animations for students or even younger children. Obviously, the image obtained seems fabricated, because it is; but we do not live in a world where technology often fabricates our vision of a real object, to the point where the brute reality appears deceptive in comparison to these images . . . but this is the case.

This book by Patrick Martinez and Alain Klotz brings together two talents: that of being accessible to all amateurs with a little perseverance, and I am willing to wager that it will be useful to a few professionals, or at the very least for students during their studies; but also the talent of not forgetting the sometimes sophisticated tools of mathematics used by professionals to process their images. Written almost without equations, this book presents powerful and accurate 'recipes' to improve the raw image obtained at the telescope's focus. Rigorously presented, these treatments will reveal to everyone the beauty of the mathematics which underlies them and might encourage some readers to deepen their understanding.

The authors, who were kind enough to do me the honor of asking me to write the foreword, no doubt expect me to express some wishes. I will give them two, for use in their next books, but also for amateurs reading this. The first is that the day soon comes when amateur observation will expand to the near infrared domain. The sky's symphony, between the blue and red of visible light, is written on a scale of wavelengths where the energy of photons varies by barely a factor of two, the exact equivalent of an octave on a piano. Include the near infrared, to which the terrestrial atmosphere is transparent, and it is an energy factor of ten that is suddenly accessible, more than three octaves. A sonata played on one octave would be horrendously mutilated if it were written for four! Now, CCDs do not recognize the distinction unique to the human eye which allows us to separate the visible from the infrared. If CCDs sensitive to infrared are still delicate to handle, since the photon energy is weaker in this spectral range, in the future I hope that many amateurs will be able to see star formations and penetrate through the hearts of the opaque clouds of our galaxy, thanks to the infrared.

My second wish is equally dear to me: professional astronomy has just conquered, not without pain, atmospheric turbulence and obtained, though not in all cases, images from the ground as sharp as those obtained from an instrument in space. This is a revolution, largely due to the marriage of the CCD and the computer. This opens up the possibility of ridding ourselves of the frustration that obliges us to assume that images obtained with a telescope of 20, 30, 50, 100 centimeters are no better than obtained with a pair of binoculars. I deeply wish that, in the shortest time possible, inexpensive adaptive systems will become available for amateurs, and I am certain of the amazing leap forward this will give our amateur observers, who will easily obtain 0.2″ images at the mediocre quality sites to which they are often confined by force of circumstances.

Thanks to CCDs and this book, amateur astronomy still has good days ahead. May I extend a warm thank-you to the authors for the pedagogical skill and rigor of their efforts.

PIERRE LÉNA

Professor at the Université Denis Diderot (Paris VII)
Researcher at the Observatoire de Paris

Preface

In less than ten years, since the early 1980s, observational astronomy has been revolutionized by the appearance of the charge couple device (CCD) detector. During this period, the large professional observatories constructed their own CCD cameras, which immediately replaced the photographic cameras in almost all areas of application.

But for amateur astronomers, doing CCD photography in the 1980s required building one's own camera, that is, mastering digital and analog electronics, computers, the science of heat . . . The situation was dire except for those whose profession gave them the necessary skills. A few pioneers, who were part of the latter group, set an example with their work and brought this new technology to the attention of amateurs. Little by little, a few groups began the adventure of constructing their own CCD camera.

By the end of the 1980s, the first commercial cameras destined for amateur astronomers made their appearance. Today, these cameras are becoming better specified and easier to use. A wider selection is available at affordable prices. It is now that we are seeing the real CCD revolution for the amateur astronomer: each will be able to use this tool and thereby increase the observational possibilities tenfold.

In 1988, the Association for the Development of Large Observing Instruments (ADAGIO) established the ambitious project of producing an 80 cm telescope geared toward amateur astronomers. It was decided, after initial research, that the principle equipment of this telescope would be a CCD camera. Because of the lack of available CCD cameras on the market that met the desired specifications, the association decided to develop the instrument themselves. Several years of teamwork and countless laboratory prototypes were necessary to achieve satisfactory competence in the field.

Just when CCD cameras are becoming accessible to all amateur astronomers, the great majority of them are still hesitant to use this new tool, because of the culture change it represents in relation to photography, but mostly because of the complexity of the first amateur cameras. The whole world is convinced that a photograph can be produced without being an optician, or developing film without being a chemist. Likewise, today it is possible to use a CCD camera without being an electronics engineer or computer expert. But the age of the pioneers is still too close for all astronomers to be persuaded.

Even if operating a camera no longer requires knowing how to build it, it is necessary to know a number of rules and procedures, in the same way that photographers learned to develop and enlarge their negatives. These procedures are simple and easy to understand and apply; but first we must find a place where we can learn.

It is therefore in the double goal of demystifying CCD cameras and teaching amateur astronomers how to use them that the ADAGIO association decided to lend its experience to arrange workshops to introduce all amateur astronomers to this new technology from 1991 on. These practical workshops, strongly oriented toward the use of CCD cameras in astronomy, enabled CCD techniques to be made available to a large number of amateur astronomers, who, for the most part, had no previous knowledge of this subject.

It was during these CCD workshops that the ADAGIO association found it necessary to publish an introductory book on CCD techniques. Naturally, it was two of the principle figures of these workshops who undertook this publication. Of course, the pedagogical experience acquired during these workshops strongly influenced the writing of this book: traditional questions asked by the organizers, their discussions, points necessitating source clarification for beginners, etc. were all taken into account.

This book is organized in a way that follows the concerns of the amateur astronomer who is discovering the world of the CCD, while searching for ways to equip himself or herself with a camera, as well as the best ways to take advantage of this revolutionary detector.

The first chapter is a presentation of material specific to CCD astronomy: What is a CCD detector and what are its capabilities? How does a CCD camera work? What role do computers play? The first chapter's goal is to give the beginner the basic ideas on this material.

The second chapter details the different characteristics of a CCD camera. These ideas are essential for the observer who hopes to understand the performance of their detector, but also for the possible buyer who will learn the criteria and arguments that will allow him or her to make a choice among the several different models available on the market. We know of amateur astronomers whose only concern for a CCD camera consists in dividing its price by its number of pixels, which is equal to buying a car based on its price divided by its length, the best car being the one that is least expensive per meter! We wanted to point out that it is important to examine several other aspects. Just as there is no perfect car, there is no best CCD camera, but there are cameras better or worse suited to the type of astronomical work being carried out.

The third chapter is specifically geared toward the astronomer who is equipped with a camera. This chapter reviews the different phases of operation of the camera and the production of images. Here, the text insists on a certain number of procedures which must be respected in order to obtain good CCD images.

The fourth and fifth chapters are concerned with the computer work carried out on the images provided by the camera. Firstly, we examine the image visualization and analysis functions, whose goal is to display and measure the data contained within the image, without modifying its content. Then, we enter into a discussion of the image treatment functions, which consist in transforming the image contents with the goal of extracting the best information possible.

Finally, chapter six examines the different possible uses of the CCD in amateur astronomy. Although this is only an introduction, it is obvious that the CCD, through its diversity, represents the tool of the future for amateur astronomers.

Even if it is more logical, it is not necessary to read this book in the exact order of its chapters. A reader who is passionate about image treatment can jump directly to chapter five; just as an observer who has a camera, but is having problems operating it, can begin at chapter three.

In order to allow this reading theme, while respecting the logical progression of the above-mentioned chapters, there is some repetition in certain chapters. But this repetition is on fundamental points that seemed interesting to present in a redundant manner.

Interspersed throughout the book, key points are presented in boxes. Their aim is to allow a quick read of the book: the hurried reader, therefore, could for a first read get an overview of the book simply by reading the boxes and looking at the figures and in this way pick up the essentials of CCD astronomy.

One of the problems with a book devoted to a technology evolving so quickly is its permanence. At the same time, even if the technology evolves, and the material rapidly becomes obsolete, the main principles will remain the same. The essence of this book is a representation of the main principles of the operation and use of CCD cameras. Several examples are taken from material existing at the time of publication, but these examples will not remain valid forever. We hope, therefore, that this book will guide amateur astronomers for several years to come.

We invite you now to enter the magical world of the CCD.

1 CCD equipment for astronomy

1.1 The CCD detector

1.1.1 *From photography to CCD*

The first observations of the sky relied on the naked eye. In this way, we can observe celestial objects to the 6th magnitude, with an angular resolution in the order of an arcminute. At the beginning of the 17th century, Galileo showed us that, with the use of an optical instrument, we can observe much fainter objects with a better resolution. Hence, a modest 20 cm telescope allows observation, visually, of 12th-magnitude stars with a resolution in the order of an arcsecond.

At the end of the 19th century, the appearance of photographic film turned our vision of the cosmos upside down. Photography, coupled with large telescopes, allowed the observation of objects of the 20th magnitude thanks to the possibility of integrating light. The general public was thus able to see for themselves superb images from the celestial world. And is this not the usual starting point for amateur astronomers?

The quality of specialized photographic films for astronomy has continued to improve, especially during the 1970s, thanks to the hypersensitization of fine-grain films. Bear in mind that the grains, whose average size is about 5 micrometers (5 thousandths of a millimeter), are the elementary points that form the photographic image.

The 1980s saw the rise of CCD cameras, which replaced photography in astronomy. CCD stands for charge-coupled device. A CCD camera takes the form of a box equipped with a transparent window inside which is located in the CCD chip. The chip consists of a mosaic of light-sensitive electronic microcells. These microcells are also called photodiodes or 'pixels', a contraction of 'picture element.' They have a rectangular or square shape about 10 micrometers on each side. The pixel mosaic is called an array. As in photography, we can take exposures lasting several minutes with a CCD since each pixel keeps the amount of light received in memory.

We can make an analogy between a whole CCD array and a photographic film. The idea of the photographic film's grain is replaced by the idea of the pixel in a CCD. In a photographic emulsion, grains have different sizes and are spread out in a random fashion, while the pixels from a CCD detector are all identical and perfectly aligned in lines and columns.

An obvious fact appears: the image formed by an optical instrument on a

FIGURE 1.1 A LYNXX CCD camera. The CCD chip can be seen as a tiny gray square behind the transparent hub.

CCD detector is automatically divided into pixels. This phenomenon is called sampling of the image. In photography, sampling also exists, but it is caused by the grains. Sampling of an image is one of the factors determining the maximum resolution attainable; it is linked to the size of the pixels in the CCD (or the photographic grain) as well as the focal length of the optical instrument.

The sensitive surface of a CCD detector is composed of a large number of juxtaposed square or rectangular elements that are sensitive to light and are called 'pixels'. Each pixel constitutes an elementary point of the image, as does each silver halide grain of a photographic film.

Photographic films must be developed by a chemical process (developer and fixer) to convert the stored light in each grain to a density of gray. This process is long (about 30 minutes) and can only be done at night's end in a darkroom. In contrast, each pixel from a CCD detector converts light into electrons. These electrons are read, converted into numeric values called ADUs, and then displayed in different levels of gray on a computer screen, all in only a few seconds. We can therefore see the image immediately after the exposure, which is very rewarding over the course of a long night!

Each pixel creates and accumulates an amount of electrical charge proportional to the amount of light it has received. Hence it is the reading of the accumulated charges in the different pixels that allows the image received by the CCD to be reconstructed.

Strictly speaking, the term 'pixel' signifies the information contained in a single point of the

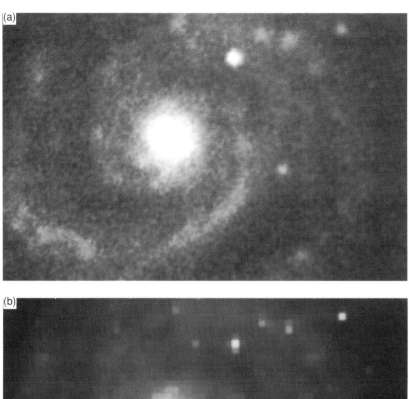

FIGURE 1.2 (a) An image of the center of the galaxy Messier 51 obtained with photo-graphic film. The image is sampled in grains. (b) The same galaxy observed with a CCD array. The image is sampled in pixels. Photographic and CCD images by A. Klotz and G. Sautot.

image, while the elementary sensitive cell of the CCD detector is given the name 'photosensor'. These two terms are often confused because each image pixel provided by a CCD array corresponds to a photosensor in that array. In any case, these two notions can sometimes correspond to different realities. For example, during an image read, electric charge can be successively transferred to several photosensors to give one single pixel. Meanwhile usage demands that we use the term 'pixel' each time it is unnecessary to distinguish the physical element of the array from the generated image point. We respect this usage throughout the book.

1.1.2 *What does a CCD detector look like?*

A CCD detector usually takes the form of an integrated electronic circuit: a gray ceramic casing with pins for electric connections (see figure 1.3). An opening on the top of the casing permits the light to access the sensitive element, the silicon array. This opening is generally sealed by a thin glass plate to protect the array. The sensitive array is clearly visible upon inspection of the circuit; it is dark in color (since its role is to absorb light) and rectangular in shape (since it is made up of a large number of pixel lines placed one beside the other). On the edge of the array, one can see tiny golden wires which are the electric connections between the array and the rest of the circuit.

1.1.3 *Geometric characteristics*

1-D and 2-D arrays In geometric terms we can distinguish two types of CCD detectors:

- 1-D arrays made up of a single line of photodiodes; these linear detectors are rarely used in astronomy today.
- 2-D arrays made up of several identical lines side by side: it is these two-dimensional detectors that are used to produce images.

It is possible to produce images with 1-D arrays, by exposing the 1-D array to one 'line' of the image and shifting it by its width to acquire the following line.

FIGURE 1.3 A Kodak KAF-0400 CCD array in its 24-pin casing. Beneath the protective window, the sensitive surface is 6.9×4.6 mm (the image shown is at a scale of 2:1).

FIGURE 1.4 A Thomson TH7805 CCD 1-D array is composed of a single line of photodiodes (the image shown is at a scale of 2:1).

This technique is used by Earth-observing satellites such as *SPOT*. The 1-D arrays are oriented perpendicularly to the satellite's motion. The orbital motion of the satellite makes the Earth images pass over the 1-D array. Each time the apparent motion of the Earth corresponds to the width of the array, a new line of the image is obtained.

In astronomy, it is necessary to use long integration times because of the faintness of the targeted objects; it would therefore be unreasonable to want to capture an image of the sky by successive exposures for several minutes on each line. It is thus preferable to use 2-D rather than 1-D arrays.

CCD arrays are made up of square or rectangular pixels, juxtaposed and evenly spread like squares on a chess board. The 1-D arrays are only formed by a single line of sensitive elements and are rarely used in astronomy.

In the rest of the book, we will concern ourselves essentially with 2-D CCD arrays; however, it should be noted that a 1-D array works similarly to a 2-D array, and almost everything that will be covered could be applied to the 1-D array.

Pixel size Pixels are rectangular surfaces measuring only a few micrometers on each side. Among the smaller sizes, for example, is the Kodak KAF-1400 with pixels of 6.8×6.8 μm, or the Loral 441 and 481 with pixels of 7.5×7.5 μm. Among the larger sizes, for example, is the Reticon RA512 with pixels of 27×27 μm, or the Thomson 7852 with pixels of 30×28 μm.

It is preferable to use an array with square pixels. When a CCD image is displayed on the computer screen, each point of the screen generally corresponds to a pixel. The points on the screen are equidistant; if this is not the case with the pixels, the image will be distorted on one of the axes.

If the pixels are almost square, the distortion is not very pronounced and can possibly be compensated by playing with the horizontal and vertical adjustments of the screen. This is the case, for example, for the LYNXX and ST4 cameras, which use the Texas Instruments CCD TC 211 as a detector. Its pixels are 16×13.75 μm. By contrast, the ST6 camera uses a Texas Instruments CCD TC 241, whose pixels are 27×11.5 μm; to avoid too big a distortion, this camera

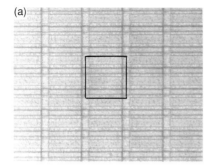
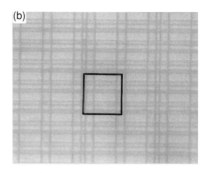

FIGURE 1.5 Photographs taken under a microscope of a pixels from a TH7895M CCD (a) and a TH7852 CCD (b). The black rectangle delineates the surface of one pixel. Each pixel of the TH7852 is 28×30 μm; note the presence of anti-blooming channels in the shape of double, dark, vertical lines while the square pixels of the TH7895M are 15 μm and do not have channels. Also note that the interior of the pixels is not homogeneous because of the grids reserved for charge transfers. Documentation: SupAéro.

automatically adds consecutive pixels two by two so that the user has the impression of working with half the number of pixels measuring 27×23 μm.

With a CCD with non-square pixels, it is still possible for the points on the screen not to correspond with the pixels while displaying the image; each point on the screen is therefore an interpolation of neighboring pixels. This is not, however, the easiest way to work. We will see, in chapter 5, how the power of image processing allows this obstacle to be avoided.

The sensitive surface and dead zone of a pixel The dimensions of pixels given by the CCD manufacturers correspond to the total size of the pixel, that is, more precisely, to the distance separating the centers of two consecutive pixels. But the sensitive area does not occupy the entire surface, inasmuch as a physical separation must exist between two consecutive pixels.

In general, this separation is very narrow and we can, in practice, consider the entire surface of the pixel to be light sensitive. There are two exceptions to this rule:

- Some CCDs like the Thomson 7852 are equipped with an anti-blooming drain. The role of this device is to discharge excessive electric charges in case a pixel is saturated by a very bright light (a star that is too bright, for example); in this way we avoid the situation where charges overload the pixel and spill into its neighbors. Physically, this drain is made up of a dead zone that borders the sensitive part of each pixel; thus, pixels from TH7852, which measure 30×28 μm, have a sensitive surface of only 30×19 μm.

FIGURE 1.6 A photograph taken under a microscope of pixels from a Sony ICX 027 BL CCD. Each pixel is 8.3×12 μm. Only the dark areas are light sensitive; the vertical light zones are reserved for interline transfer. Documentation: Jean Dijon.

- Some CCDs operate on an interline transfer: this means that a zone insensitive to light is reserved between each line to store the electric charges away from the light between the end of the exposure and the moment they will be read. This is the case, for example, with the Sony CCD; this manufacturer, which gives little information on its products, does not specify the size of this dead zone; it seems to have relatively small dimensions (see figure 1.5b).

The presence of a dead zone of non-negligible size between the sensitive surfaces of consecutive pixels corresponds to a loss in sensitivity in the CCD in proportion to these areas. This is not a major drawback. But in astronomy, dead zones pose other problems, essentially because of the small size of the stellar images: if the telescope provides very sharp images of stars, smaller than the size of the dead zone, the stars that, by chance, fall directly into a dead zone could simply go unnoticed! In practice, images are always spread out by diffraction, optical aberrations, atmospheric turbulence, or a poor focus; it is therefore probable that an image of a star will overrun and appear on the adjacent pixels. Nonetheless, an important part of its light is lost, and it is then impossible to proceed with precise photometric measurements, although this can be improved by deliberate defocussing.

The sensitive parts of the pixels of a CCD are generally joined; the entire surface of the array, therefore is light sensitive. Nevertheless, certain types of arrays possess dead zones, which are not sensitive to light, between each pixel. These dead zones obviously reduce the entire sensitivity of the array. They are of little consequence for images of extended objects but can cause errors in stellar photometry.

Color CCDs and black and white CCDs A CCD detector's pixels receive light and transform the amount received into an electric charge. If all the pixels are identical and receive light from the entire spectrum, the CCD provides an image in which colors cannot be differentiated; it is a black and white image. To obtain a color image, we must use a CCD which has one of its three pixels equipped with a red filter, the second with a green filter, and the third with a blue filter. During a read, we consider that these three pixels have the three fundamental colors of the same point on the image. This principle is used in video cameras equipped with CCDs and operates very well on everyday objects that have a large enough surface to illuminate each of the three pixels evenly.

But in astronomy, what has been said in the previous paragraph about dead zones and the sharpness of stellar images applies here too: imagine the results of a star field with some stars only focused on the red pixels, others on the green pixels, and others on the blue pixels!

CCDs used in astronomy are all black and white detectors. It is possible, however, to produce color images by trichromatism: the observer must take three images of the same object, each one with one of the three fundamental filters, and then combine these images during analysis.

The number of pixels and the size of the array The number of pixels of an array is the product of the number of pixels per line by the number of lines. It is an important criterion for the power of the array because the field of the image obtained depends on them. The more pixels there are, the longer the read time of the image will be, the larger the computer's memory has to be, and the more expensive the CCD!

The Texas Instruments TC 211 (192×165 pixels) and the Thomson 7852 (208×144 pixels) are two of the smallest CCD detectors used in astronomy; they have the advantage of a very reasonable price (less than £140/$200 a piece). A 512×512 pixel CCD is a very comfortable size for an amateur astronomer but its price is close to £1400/$2000. Several manufactures offer 1024×1024 pixel CCDs or even 2048×2048 pixels (with corresponding prices!). Prototypes with 4096×4096 pixels are now appearing. The evolution is very fast in this field: every year larger and larger CCDs appear, while the prices for the larger arrays come down, as all the while more high-performance models are being introduced.

To know the sensitive surface size of a CCD, it suffices to multiply the number of pixels by the size of the pixels. Therefore, the surface of the TC 211 (192×165 pixels 13.75×16 μm in size) is 2.64×2.64 mm; the Thomson 7863 (384×288 pixels 23×23 μm in size) is 8.8×6.6 mm.

We can see that CCD arrays are smaller in comparison with the standard formats of photography (24×36 mm); only the very large arrays of 2048×2048 and 4096×4096 reach a surface comparable to 24×36. This is the main area where photographic film still has an advantage over CCDs. Today, the CCD has replaced photography in all areas of professional astronomy, except when the field to be covered is important, for example, observations done with Schmidt cameras.

FIGURE 1.7 Two CCD components of very different sizes (shown at a scale of 2:1): the Thomson TH7895 contains 262 144 pixels and is 1 cm^2 in size. The Texas Instruments TC211 only has 31 680 pixels and is just 0.07 cm^2 in size.

1.1.4 *The operation of the CCD*

The reading of the CCD The CCD package is equipped with three types of electric connections:

- some pins receive, from the camera's electronics, continuous voltages (for example, 1.5 V, 12.8 V, and 15 V for the Thomson 7863 array) which electrically feed the different CCD components.
- Some pins receive, from the camera's electronics, voltages that vary between a base value and a high value (for example, between 0.3 V and 10 V for the same array). These signals are called clocks.
- An output pin allows the reading of electric charges stored in each pixel.

The reading of a CCD array consists in moving charges from different pixels.

These movements are caused by changes in the levels (high→low or low→high) of the signal clocks. To each type of movement there corresponds a very precise sequence of changes in level of the clocks, these sequences are specified by the CCD manufacturer in a schematic called the timing diagram.

CCD detectors possess, besides the pixel array, a line of CCD cells called the horizontal register or line register. This register contains as many cells as one line in an array, but these cells are masked so as not to be light sensitive.

CCD array

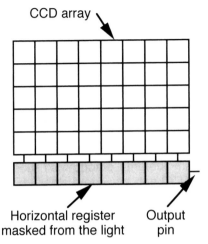

Horizontal register   Output
masked from the light   pin

FIGURE 1.8 A schematic representation of a pixel array (here 8×5 pixels) as well as a CCD camera's horizontal register. The horizontal register is masked from the light (shaded gray in the figure). Its role is to receive the array's electric charges and transport them to the CCD's output pin.

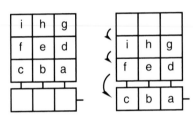

FIGURE 1.9 The workings of the horizontal clocks: the electric charges of the pixels of each line of the array are transferred to the corresponding pixels immediately underneath; the last line (a, b, c) is transferred into the horizontal register.

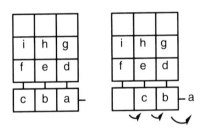

FIGURE 1.10 The workings of the horizontal clocks: the electric charges of each pixel of the horizontal register are transferred to the neighboring pixels and the last pixel (a) is transferred to the exit.

When we operate on certain clocks called the vertical clocks, each CCD line is transferred to the line directly underneath it, and the lowest line is transferred into the horizontal register (see figure 1.9).

When we operate on certain clocks called the horizontal clocks, each pixel's electric charges are laterally transferred to the neighboring pixel, and the last pixel is transferred into an output amplifier (see figure 1.9); the CCD's charge sensing gate therefore takes a voltage dependant on the quantity of electric charges contained in the exiting pixel.

The reading of a CCD array, therefore, consists in executing the following sequence: the electronics associated with the camera sends a vertical clock signal to transport a line of pixels into the horizontal register. Then, a horizontal clock cycle is sent to transport the pixel to the register's extremity for the output stage; hence, it is possible to read the electric charge's value contained in the pixel. This last operation is repeated as many times as there are pixels in the line. Once the entire line is read, the camera's electronics send a new vertical clock signal to transport the next line into the horizontal register, and so on until all of the lines have been read.

The electronics connected to the CCD's output pin sees the values of the electric charges contained in the array's pixels file by, one by one.

The integration of an image If no clocks are activated, no pixels are transferred. In this state, any light on the pixels generates electrons that are stored in place and form the electronic image corresponding to the light received. Making an exposure on a CCD consists, therefore, in leaving the clocks inactive.

Once the integration time of the image is finished, all of the pixels are transferred to the output, as previously described. Each photosensor, then, is emptied of its charge and is ready to receive light for the next image.

This permanent light sensitivity of the photosensors results in charges that are generated and stored during the time before we decide to begin the exposure (during the aiming of the telescope, for example). Even if, during this phase, the CCD is placed in darkness, electric charges called thermal charges are spontaneously generated and pollute the charges created by the subsequent image. So, prior to beginning an exposure, we must 'clean' the CCD; this is done by simply transferring all of the pixels to the output (several times for security): the camera's electronics sends the clock cycles corresponding to an exposure read, but at a much faster speed so as not to waste too much time and not to have to worry about measuring the values at the output stage.

Image integration begins right after CCD cleaning and ends with the charges being transferred toward the output in the array reading procedure.

The operations described so far are those of an array with all of its lines exposed to light and participating in the image. These arrays are called *full frame*. The problem with these arrays is that during the read-out, the electric charges stay exposed to light in the photodiodes to which they are being successively transferred, while waiting their turn to enter the horizontal register (which is not exposed to light).

Unfortunately, the read time for an array is not negligible being from one to a few seconds. If we have just exposed a CCD for 20 minutes on a galaxy field of 14th magnitude, these few seconds of image transfer should not leave a trace;

FIGURE 1.11 This image of Jupiter shows the smearing effect characteristic of the ST4 CCD camera, which does not have a mechanical shutter. This image has been specially treated to enhance this phenomenon. Image obtained by Xavier Palau, Barcelona.

but if we were reading a planetary image that we had exposed for 1/10 of a second, the planet would induce an unwanted streaking on all the pixels that pass across its image during the transfer. This streaking phenomenon is called *smearing*.

To avoid smearing, full frame CCD cameras should be equipped with an electromagnetically controlled shutter, released by the camera's electronics. The shutter is closed once the exposure time has ended and in this way protects the array from light during the read-out.

Other CCD arrays are composed of two zones containing the same number of lines: one zone called the image zone, where the pixels normally receive light, and a second zone masked by a screen, called the memory zone. At the end of the integration time, the camera's electronics sets off the vertical clocks as many times as there are lines in each zone, in order to transfer all of the pixels in the image zone into the memory zone, where they are shielded from the light. This transfer is much faster than the reading of the array since it is not necessary to stop between each vertical clock cycle to read the pixels in the horizontal register. Hence we can take our time to read the memory zone. These arrays are *frame transfer* arrays and some manufacturers describe this operation as an electronic shutter.

In most cases frame transfer arrays allow a mechanical shutter to be dispensed with. However, we can consider them less effective in certain cases, such as planetary imaging, since the transfer time from image zone to memory zone cannot be completely negligible compared with the integration time in this case (about 2 ms to transfer 100 lines).

Some cameras, equipped with a full frame CCD, but devoid of a shutter, offer a mode of functioning based on frame transfer: at the end of the integration time, the top half of the image's photosensor charges are quickly transferred to the bottom half, so as not to delay the transfer from the upper half. Hence, the lower half of the image is lost, but it becomes possible to make a planetary image on condition that it only occupies the upper half of the array. This mode is called half-frame transfer. This clever function mode implies that no bright

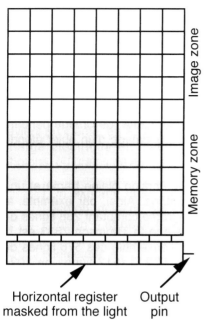

Image zone

Memory zone

Horizontal register masked from the light

Output pin

FIGURE 1.12 A schematic representation of a frame transfer CCD array (here 8×5 pixels). The image is formed on the photosensors in the image zone (white). At the end of the exposure, all of the electric charges are quickly transferred into the memory zone, which is shielded from light (shaded gray), this allows us to bypass a mechanical shutter. The memory zone has the same number of photo-sensors as the image zone. Beneath the memory zone is the horizontal register, also masked from the light, its role is to receive the electric charges from the memory zone and bring them to the CCD's output pin.

FIGURE 1.13 The TH7863 CCD component has a memory zone (light area) reserved for frame transfer at the end of the exposure. Thus, we can do without a mechanical shutter in most cases.

objects (stars, planetary satellites) occupy the lower half of the array (its light would create charges in the pixels of the upper part of the image during reading); it cannot be applied to images of large objects, such as the Moon or Sun, in particular.

Some manufacturers offer full frame and frame transfer versions of the same array. Thus, the Thomson 7863 is a frame transfer array containing 288 lines of 384 pixels in each of its zones, whereas the Thomson 7883 is the same array but the image zone contains the full 576 lines of 384 pixels. Also, the Texas Instruments TC213 (512 lines of 1024 pixels) is the frame transfer version of the TC215 (1024 lines of 1024 pixels). The large arrays are all full frame technology, partly because the frame transfer time becomes too long, and partly because their price easily justifies the purchase of a shutter rather than wasting half the array's surface simply for memory.

Another type of array offers a memory line beside each sensitive line. These arrays are called *interline transfer* arrays. They do not need an electromagnetic shutter. Their drawback is that the memory line is essentially a dead zone between each pixel line of the image. The ratio of sensitive surface to total surface of a CCD array is called the fill factor. Some Kodak and Sony arrays are equipped with interline transfer (see figure 1.6). In this case, the electronic shutter is particularly quick and efficient, since only one clock cycle transfers each sensitive line to its shielded neighbor.

Full frame arrays are light sensitive on their entire surface; they generally need a mechanical shutter. Frame transfer and interline transfer arrays have a memory zone where the pixel charges are shielded from the light during the reading time.

1.1.5 *CCD performance*

The sensitivity of a CCD The CCD's sensitivity is expressed as electric charges per unit of light energy received, hence in coulombs per joule, or, what turns out to be the same, amperes per watt. But we can equally measure electric charge in electrons and the quantity of light in photons.

A CCD's sensitivity is expressed as the number of electrons produced per incident photon, which is an easy concept to use. This idea is called the *quantum efficiency* (QE); it is very useful for characterizing the effectiveness of a detector.

If the CCD were a perfect detector, it would produce one electron each time it received one photon (yielding 1, or 100%). A CCD reaches a QE of about 50% at certain wavelengths (1 electron produced for 2 incident photons), which is remarkable; in comparison, the best photographic films have a QE of 4–5%.

Note, in figure 1.14, that CCDs are, for the most part, sensitive from 400 nm (the violet limit of visible light) to 1 μm (the near infrared). The maximum sensitivity is situated in the red, around 600–700 nm, which is fortunate for observations in the Hα line of hydrogen (635.5 nm).

For certain observations, it is hoped that the violet sensitivity can be improved and even extended into the near ultraviolet. Three routes are possible:

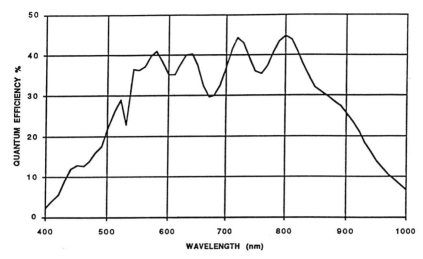

FIGURE 1.14 An example of the spectral sensitivity of a CCD: the Kodak KAF0400. Compare this curve to that of the Thomson component shown in figure 1.15.

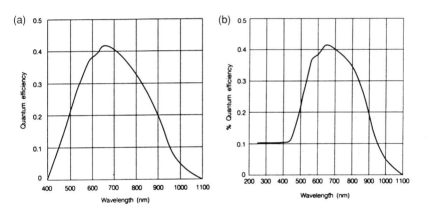

FIGURE 1.15 The influence of a 0.4 to 0.6 μm thick fluorescent coat deposited on to a Thomson 7895 CCD. To the left, the spectral sensitivity of an untreated CCD; to the right, the sensitivity of a treated CCD. From the documentation of THOMSON-CSF-Specific Semi-Conductors.

- Cover the CCD with a thin fluorescent colored coat which, excited by ultraviolet light, re-emits in the green, a color to which the CCD is sensitive (see figure 1.15).

- Use a *back-lit CCD*; this means that the light no longer arrives from the electrode and insulator side but from the silicon side. The silicon layer must therefore be very thin so that the CCD operates under these

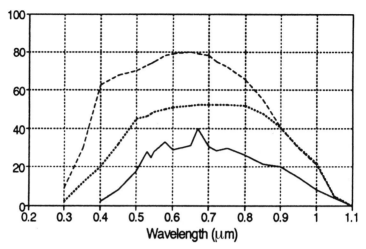

FIGURE 1.16 Comparison of the spectral sensitivity of Scientific Imaging Technologies
Inc.'s TK1024 CCD in a version illuminated from the front (solid line), a thinned
version (dotted line), and a thinned version covered with an anti-reflective coating
(broken line). Documentation; Document SITe.

conditions, from which comes the name *thinned CCD*. The gain in
sensitivity is important, especially in the shorter wavelengths (see
figure 1.16), but a thinned CCD presents other inconveniences, such as
the appearance of monochromatic light interference fringes, especially
in the near infrared. For the amateur astronomer, the main drawback
of the thinned CCD is a generally unattractive price (it being about 10
times more expensive than the same CCD not thinned!) linked to the
difficult manufacturing of the product.

- Use an interline transfer CCD, which is usually more blue sensitive.

The capacity of a pixel A pixel's electron capacity is not unlimited. The capacity
depends on the CCD: as a general rule, it is larger for larger pixels; furthermore, MPP technology CCDs (which we will see in the context of dark
current) have a capacity two to four times smaller than traditional technology
CCDs. The values given by the manufacturers go from 45 000 electrons for the
Kodak KAF1400 (6.8×6.8 μm pixels with MPP technology) to 940 000 electrons for the Thomson 7863 and 7883 (23×23 μm pixels with traditional
technology).

Once the amount of light that hits a pixel during the integration time is
high enough, the number of electrons generated can exceed the pixel's capacity. Hence, it becomes saturated, and the excess pixels have a habit of spilling
into neighboring pixels, preferentially those in the same column. Moreover,

FIGURE 1.17 An image of the center of Messier 42, obtained by a LYNXX CCD (Texas Instruments TC211 detector) at the focus of a 500 mm photographic camera open at $F/D=3.5$. The exposure is 60 seconds, which is sufficient to provoke a saturation of the Trapezium stars; notice the blooming trails, characteristic of saturation. Image by: P. Martinez and P.M. Berger.

during the reading time, the total charges from a saturated pixel cannot simply be emptied at once: residual charges therefore pollute the following pixels of the same column that pass by the saturated site during the reading transfer.

This phenomenon often occurs in astronomy when a bright star is in an image field during a long exposure: the extra electrons saturate not only the pixels that form the stellar image but some neighboring pixels in the same column, usually in the direction of the charge transfer during the array read.

In practice, once a bright star saturates pixels, we notice a characteristic white tracking beginning from the star (see figure 1.17); this phenomenon is called *blooming*. To avoid, or at least reduce, the blooming trails, some CCDs are equipped with anti-blooming systems, which allow the emptying of excess electrons during integration. Unfortunately, the anti-blooming device sometimes introduces other problems: the existence of a dead zone in the photodiode linked to the draining electrode and the loss of linear response for very high illuminations.

In fact, the inconvenience of blooming is usually an aesthetic one. It is rare, in fact, for the faint object we wish to detect to be located in the proximity of a bright star; and if this is the case, we can usually set up the camera so that the saturation trail does not interfere with the studied object.

Pixel capacity is not the sole criterion of choice for a CCD array. One must always compare this capacity with the noise value (the random fluctuation around the average signal level) of the provided image. For example, an array whose pixels contain 100 000 electrons with a noise of 10 electrons has the same dynamic range as an array whose pixels contain 500 000 electrons but with a noise of 50 electrons. Note that in these two cases the image's dynamic range is 10 000, which is excellent compared to the performance of photographic films. Also, notice that the first array would be better adapted to faint objects (lower noise, therefore better detectivity) while the second array would be preferable for planetary imaging (minor influence of photon noise on a larger capacity).

Finally, let us address a tricky problem sometimes encountered with front-lit CCDs: remnant image. This occurs when a highly cooled CCD is exposed to a very strong near-infrared flux (at wavelengths larger than 800 nm). The photons penetrate very deeply into the array and stay trapped for several minutes, even if the array is emptied several times. They finally produce an additional 'ghost signal' on the later exposures. In practice, this phenomenon is only observed when the CCD is mounted behind a spectrograph during the calibration exposures with emission lamps.

The linearity of a CCD The CCD is essentially a nearly perfect linear detector. This means that the number of electrons generated in one pixel is proportional to the amount of light that hits that pixel. The linearity is usually better than 0.01%, which is excellent. But, of course, this rule no longer holds once the CCD is saturated.

This linearity gives the CCD three very important advantages for astronomy:

- Firstly, the CCD has practically no detection threshold. An object which has a very faint luminosity will still generate electrons, albeit very few. If this number of electrons is less than the image noise, the object will not appear on an image, but we can produce a large number of images and combine them. The result of this operation is to average the noise while the small quantities of electrons created in each image by the object combine to appear above the noise. This is only rarely possible with photographic film, since a faint object, beneath the film's sensitivity threshold, does not provoke any reaction in the film's emulsion; therefore, combining a large number of images that contain nothing is useless.

- The number of electrons generated is proportional to the quantity of incident light, regardless of the integration time. Unlike film, the CCD is not affected by the Schwarzschild effect or deviation from the law of reciprocity. Exposing 1000 seconds on a CCD generates exactly ten times more electrons than a 100 second exposure would; unfortunately, this is not the case with photographic film, which undergoes an important drop in sensitivity as the exposure time is lengthened, making long integration times almost useless.

- Since the number of electrons generated is proportional to the quantity of light received, the reading of a CCD array directly leads to a measurement of the luminosity of all the objects that constitute the image; the application in photometry is immediate and of great precision (a few hundredths of a magnitude if done properly). In photographic photometry, the density of the film on the star's image must be measured and compared to the film's calibration curve (produced under very precise conditions) to deduce the quantity of incident light; the work is more tedious and much less accurate. The detector's linearity gives the CCD another important advantage over film when it comes to photometry of extended objects: in photography, it is very difficult and imprecise to compare the luminosity of a comet, whose image is spread out, with the point-like images of reference stars; with CCD, it is sufficient to add, on the one hand, the signal of all of the pixels that constitute the comet's image, and on the other, the signal of all of the pixels that constitute the reference (comparison) star's image.

A CCD's linearity gives it, in astronomy, three important advantages over photographic film:

- The CCD has practically no detection threshold.
- The schwarzschild effect is unknown for CCD.
- The number of electrons generated gives a direct photometric measure.

Transfer efficiency During the CCD read, the charges move from cell to cell along the array's column, then into the horizontal register until their exit. Unfortunately, these transfers are not 100% effective: a small percentage of transferred electrons are lost at each transfer. This phenomenon has two ill-fated consequences. Firstly, the quantity of electrons read in a strongly illuminated pixel is slightly less than the quantity of electrons generated because of electrons lost en route. Secondly, these lost electrons are recovered by subsequent pixels, which augment the charge.

The inefficiency of the transfer, however, is very small: CCD manufacturers often claim a transfer effectiveness in the order of 0.99999, which means that only one out of 100 000 electrons is lost at each pixel transfer. Thomson even claims an effectiveness of 0.999 998 (2 out of one million electrons are lost at each transfer) for its 512×512 and 1024×1024 arrays.

Of course, it is necessary that the transfer's effectiveness be better as the array's size increases, since the number of transfers for the furthest pixel is greater. For example, the Texas Instruments TC211 array (192×165 pixels), which has a transfer effectiveness of 0.99998, loses 7 out of 1000 electrons during the transfer of its furthest pixel, while the furthest pixels on the Thomson 512×512 and 1024×1024 pixel arrays lose 2 out of 1000 electrons and 4 out of 1000 electrons respectively.

From a mathematical point of view, the random fluctuations of the read electronic charge, because of the transfer's ineffectiveness, are expressed in the form of *transfer noise* expressed as

$$\sigma_e = \sqrt{\epsilon n N},$$

where ϵ is the ineffectiveness of the transfer (or $1 - $ transfer efficiency), n is the number of transfers made to read the pixel concerned, and N is the number of transported charges. If we take the preceding examples, with a transported charge of 20 000 electrons and an average number of transfers of 180 for the TC211 and 512 for the Thomson 7895, we find transfer noises of 8.5 electrons for the TC211 and 4.5 electrons for the TH7895.

Finally, note that the transfer efficiency strongly depends on the way the array is read: imperfect clock shapes or synchronization, or too fast a read, alter the transfer efficiency.

The dark current Even when the surface of the CCD does not receive any light, electrons are spontaneously generated in the photosensors and contribute to what we call dark current. This dark current also exists, of course, when an image is produced; it hampers the CCD read because it is impossible to distinguish those electrons which have no astronomical significance from those generated by the incoming light.

The dark current has certain characteristics which allow us to limit its nuisance:

• It is a perfectly reproducible phenomenon: in identical temperature conditions, a given photosensor always generates the same number of electrons per unit time with a narrow statistical dispersion which is linked to the probabilistic character of the phenomenon (but this number of electrons varies from one sensor to another).

• The generated electric charge is quasi-proportional to the integration time.

• The dark current strongly depends on the CCD's temperature: its intensity decreases by a factor of 2 each time the CCD's temperature is lowered by 6 °C (which is equivalent to saying that this intensity decreases by a factor of 10 each time the CCD's temperature is lowered by 20 °C). Because of this dependence on temperature, the dark current is also called the thermal current and the quantity of electrons accumulated represent the *thermal charges*.

All images of the sky, therefore, contain electrons created by light that contains useful information and thermal electrons that we want to eliminate. Since this thermal phenomenon is reproducible, all that is required is to produce a second

FIGURE 1.18 A thermal image obtained in darkness with a LYNXX CCD camera at ambient temperature and an exposure of 0.01 seconds. At 20 °C, the thermal charges generated during the array's read time (about 2 seconds) are not negligible and further we can see from the gradient on the image that the last pixels read are more affected than the first ones. Notice the horizontal register located at the bottom of the image.

CCD image in the dark (the shutter stays closed) with identical thermal and exposure time conditions. This second image contains only thermal electrons; it is sufficient, then, to remove it from the first one so that it only contains photon-electrons. With proper software, it is even possible to subtract a dark image produced with different temperatures and exposure times (see section 5.2.2.).

Despite the above, the thermal electron problem is not solved. Even if we know how to remove them from an image, it is necessary to limit their numbers for two reasons:

- An exposure with too high a temperature or too long an exposure would cause saturation in the pixels owing to dark current; hence, there would be no space for the data one wishes to collect.

- Even if we do not attain saturation, the dark current generates a noise called thermal noise. In fact, the creation of thermal electrons being a random phenomenon, their numbers obey a dispersion of $\sqrt{N}$. This means that if a pixel, under certain conditions, generates on average N electrons, different measurements done in these conditions will not always give N electrons, but a number that deviates on average by $\sqrt{N}$. Therefore, when we remove the dark current from an image, we know the average value generated by each photosensor, but not the random deviation from this mean occurring in the perpendicular pixel concerned. From this there results an uncertainty (or noise) of $\sqrt{N}$ in the numbers of electrons to be removed, and hence in the number of electrons attributed to photons. For example, if each pixel generates 90 electrons per second and the integration time is 40 seconds, the number of thermal electrons to remove will be $40\times90=3600$; but the

FIGURE 1.19 A thermal image obtained in darkness with a LYNXX CCD camera (the CCD is a TC211) cooled to -35 °C and a 600 second exposure; notice that not all the pixels generate exactly the same number of thermal electrons. Also notice, in the lower right-hand corner, the effect of the electroluminescence of the output amplifier (see figure 1.20).

uncertainty of this number is $\sqrt{3600}=60$. The image, therefore, would contain a thermal noise of 60 electrons, making the detection of stars, producing a smaller number of charges than that value, impossible.

Thus, it is necessary to limit the generation of thermal electrons. With this in mind, the CCD is cooled. On the other hand, a new CCD technology called *MPP* (for multi-pinned phase) allows the reduction of the dark current by a factor of about 30–50. MPP technology CCDs are characterized by a negative voltage on certain clocks.

The dark current's value is given by some manufacturers in terms of the number of electrons generated per pixel per second, for a given CCD temperature. One of the highest performance CCDs at this level (as of 1994), is the Thomson 7895, with MPP technology, which is given as 0.5 electrons per second per pixel at -40 °C.

Other manufacturers give the dark current's value in pA/cm^2. It is up to you to divide the pico-amperes (10^{-12} A) by the electron's charge (1.6×10^{-19} C), and multiply everything by the surface of a pixel in cm^2. One finds that a current of 1 pA/cm^2 corresponds to 6 electrons per second per pixel for 10×10 μm. The numbers supplied by the manufactures range from 10 to 500 pA/cm^2, at ambient temperature.

Even in the absence of all lighting, the CCD spontaneously generates a 'dark current' proportional to the integration time and exponentially increasing in relation to the temperature. Dark current has two annoying consequences:

- It spontaneously generates 'thermal noise', which limits the detection of faint stars on the image.
- It can, at the limit, saturate the photodiodes all by itself, making an image read impossible.

FIGURE 1.20 This image, exposed for 5 minutes with an ST4 cooled camera, was produced in total obscurity. Notice, in the lower right-hand corner, the signal caused by electroluminescence. It is this phenomenon that limits the camera's exposure time.

Electroluminescence On certain CCDs, the output amplifier emits a very weak light, sufficient to create charge in pixels at the array's corner near the amplifier (the corner of the first pixel read in the first line read). Like thermal electrons, these charges are subtracted once the dark image has been removed, but they generate a noise of the same value and can bring about a saturation of the pixels concerned.

This phenomenon does not appear on frame transfer arrays because the memory zone, located between the output amplifier and the image zone, protects it during integration. Moreover, the memory zone's charges are evacuated before the image zone's transfer.

Defective pixels Defective pixels are either pixels whose gain differs by a certain percentage from the array's average, hot pixels that saturate very quickly or dead pixels that stay black regardless of the light they receive. It is very difficult to produce an array free of faults, especially if it is large in dimension. Manufacturers also offer arrays of different quality classes; each class is defined by a maximum number of defective pixels. It is often specified whether the defects are isolated pixels, pixel groups, or columns; also, a distinction is often made between the array's central area, considered the most important, and the peripheral area.

1.2 CCD camera electronics

We have seen that a CCD camera's readings consist in sending a certain number of electronic signals called 'clocks' to the chip. At each clock cycle applied to the horizontal register, the CCD's output pin takes on a voltage that is proportional to the electrical charges contained in the transferred pixel.

To operate the CCD camera, we must build an electronic system that on the one hand furnishes the necessary clocks, and on the other, reads the output voltage and transforms it into interpretable information in the form of a digital image.

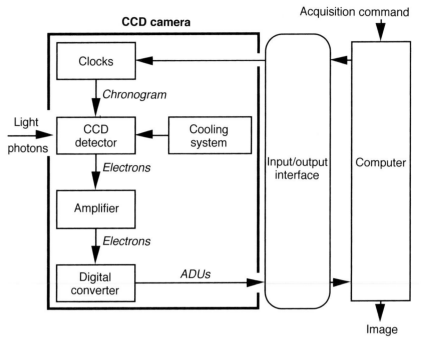

FIGURE 1.21 A schematic of the elements that control the output of images from a CCD camera. The type of information conveyed between each element is indicated in italics.

1.2.1 *Clock generation*

The definition of a clock Each type of CCD chip needs a different set of specific clocks. The variation of clock signals over time is defined in a document called the timing diagram, which is supplied by the CCD manufacturer. The manufacturer also defines the high and low voltages of the clocks as well as the continuous voltage to apply to the CCD.

For example, here we list the timing diagrams of the Thomson 7895 CCD. We have not listed all of the information necessary to build a CCD camera, however; those interested should refer to the complete Thomson documentation.

The timing diagram of figure 1.22 shows the general operation of the clocks when producing an image; this production is done in three phases: the cleaning of the CCD, integration, and reading of the pixels. The clocks Φ_{P1}, Φ_{P2}, Φ_{P3}, and Φ_{P4} concern themselves with the transfer to the horizontal register; clocks Φ_{L1}, Φ_{L2}, and Φ_{R1} pilot the horizontal register and the output amplifier. V_{OS1} is the voltage that appears on the output pin.

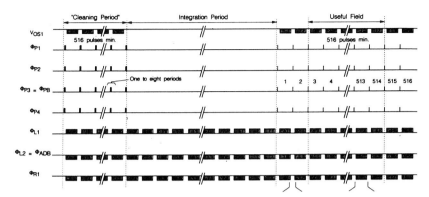

FIGURE 1.22 General timing diagram of Thomson 7895M CCD clocks. Taken from
THOMSON-CSF – Specific Semi-Conductors.

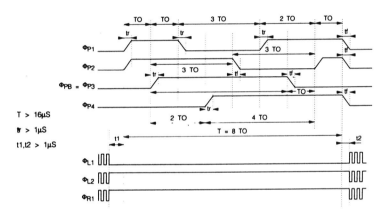

FIGURE 1.23 A detailed chronogram of ΦP clocks from the Thomson 7895M CCD.
Taken from THOMSON-CSF – Specific Semi-Conductors.

The timing diagram of figure 1.23 details the workings of clock Φ_p. The cycle
shown has the function of shifting all of the lines one row toward the horizontal
register. This cycle, therefore, is meshed between two series of cycles Φ_L, whose
function is to read the pixels of one line.

Figure 1.24 shows the horizontal register being read by the Φ_L clocks. Each
cycle represents the reading of one pixel.

One of the elementary signals is represented in figure 1.25. Note that
under the motion of clocks Φ_{L1}, Φ_{L2}, and Φ_{R1}, the output voltage takes three
successive values during the reading of one pixel: a voltage reset to zero, a
reference level whose voltage is that of the pixel if it did not contain electrons (floating diffusion), and a measuring level (video signal) whose differ-

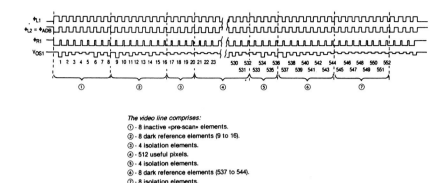

The video line comprises:
① - 8 inactive «pre-scan» elements.
② - 8 dark reference elements (9 to 16).
③ - 4 isolation elements.
④ - 512 useful pixels.
⑤ - 4 isolation elements.
⑥ - 8 dark reference elements (537 to 544).
⑦ - 8 isolation elements.
All 40 dummy elements are read during the line blanking period.

FIGURE 1.24 A detailed timing diagram of ΦL clocks from the Thomson 7895M CCD. Taken from THOMSON-CSF – Specific Semi-Conductors.

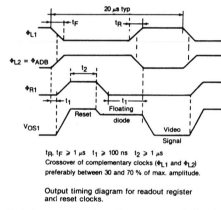

$t_R, t_F \geqslant 1\ \mu s$ $t_1 \geqslant 100\ ns$ $t_2 \geqslant 1\ \mu s$
Crossover of complementary clocks (Φ_{L1} and Φ_{L2}) preferably between 30 and 70 % of max. amplitude.

Output timing diagram for readout register and reset clocks.

FIGURE 1.25 A detailed chronogram of an elementary cycle for the clocks of the Thomson 7895M CCD. Taken from THOMSON-CSF – Specific Semi-Conductors.

ence in voltage from the reference level corresponds to the electric charge contained in the pixel. The output signal's appearance is very important for measuring the charges; we will come back to it in our discussion on reading electronics.

The extract from the technical documentation of TH7895M shown below defines the voltages to apply to each of the following clocks whether they are at a high level or low level. Notice how the voltages for the line transfer clocks take negative values, a characteristic of MPP technology.

Table 1.1, also an excerpt from the technical documentation of the TH7895M, defines the continuous voltage to be applied to the power supply connectors of the CCD.

Table 1.1

Parameter	Symbol	Value Min.	Value Typ.	Value Max.	Unit
Image zone clocks during integration period	Φ_{P1}, Φ_{P2} Φ_{P3}, Φ_{P4}	-11	-10	-8	V
Image zone clocks during transfer period	(Φ_P) Low	-11	-10	-8	
	(Φ_P) High	0	$+2$	$+2.5$	V
Image-to-register clock	(Φ_{PB}) Low	-11	-10	-8	
	(Φ_{PB}) High	0	$+2$	$+2.5$	V
Output register, reset and summing well clocks	$(\Phi_L$, Φ_{R1}, $\Phi_{ADB})$ Low	0.0	0.3	0.8	V
	$(\Phi_L$, Φ_{R1}, $\Phi_{ADB})$ High	10	11	12	V

Table 1.2

Parameter	Symbol	Value Min.	Value Typ.	Value Max.	Unit	Remark
Output amplifier drain supply	V_{DD}	18	19	19.5	V	
Reset bias	V_{DR1}	12	12.5	13	V	
Substrate bias	$V_{SSA.B}$		0.0		V	Note 1
Output load resistor	R_1		47		kΩ	Note 2
Register output gate bias	V_{GS1}	1.3	1.5	1.7	V	
Temperature sensors	V_{TA}, V_{TB}		15		V	
Bias 1	V_1	14	15	16	V	
Bias 2	V_2	8	10	12	V	
Bias 3	V_3		V_{SS}		V	
Bias 4	V_4	-11	-10	-8	V	

Note 1: $V_{SS}=0$ V requires that the register and reset clocks, loaded by the device, do not contain spurious negative spikes (less than -0.1 V).

Note 2: At low output frequency, the load resistor is increased in order to lower power consumption and noise (resulting from amplifier bandwidth limitation).

Read time and window mode On the timing diagram shown in figure 1.25, Thomson suggests that a pixel read cycle last 20 μs. This length is characteristic of CCDs used in astronomy since it nears the optimum value: a shorter duration would result in a bad transfer of electric charges, thus leading to a higher noise: a long time leads to a prohibitive read time for the array. A minimum time per pixel can equally be imposed by the reading electronics (band width of filters and amplifiers, delays in conversion); this time can be more than 20 μs and possibly reach 50 μs.

The 7895 Thomson array we are using as an example contains 512 lines of 512 pixels, or a total of 262 144 pixels. In taking 20 μs per pixel, we come to a read time of 5 to 6 seconds for the complete image. This time is negligible if we are reading a faint object exposed for 10 minutes; but it would be penalizing for some work, for example, during the focusing on bright stars.

Fortunately, the jobs that need a large image rate are generally satisfied with the acquisition of a tiny portion of the sky: the image of a star for target acquisition or to produce fast photometry in the case of an occultation by an asteroid. Hence, it is enough to simply read the pixels contained in a little window of the image, from where the term windowing comes from.

In windowing mode, the non-read pixels, meanwhile, must be transferred toward the output, but the time dedicated to their transfer can be considerably reduced. The gain in read time of an image can be very significant and, of course, depends on the size of the window we decide to read.

Windowing is a function mode specific to the clocks: we must transfer at a high rate the lines that precede the first interesting line; and in this package of interesting lines, quickly transfer the non-read pixels, acquire the pixels from the window in normal read and quickly transfer the remaining pixels, so as to empty the lines beyond the window at a high rate. This implies a certain degree of subtlety in the command electronics, especially if we want the user to be able to decide what size and where the window will be located in the image.

Windowing consists in reading only part of the array that contains useful information. For example, a small square centered on a star to do the focus. The goal is to reduce the read time, therefore increasing the image rate.

Binning Up until now, we have described the term 'pixel' as being the elementary site of the CCD array. In reality, it would be fairer to describe the elementary sites by the term *'photosensor'*. Usually, each image's pixel that reaches the computer corresponds to a single photosensor on the CCD detector. Therefore, it is normal to abuse the term pixels when describing a CCD array's

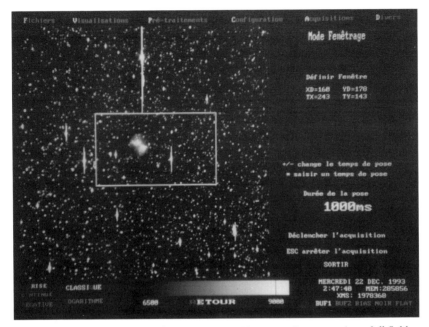

FIGURE 1.26 An image provided by an ALPHA 500 camera. In automatic, at full field, an image is formed every 6 seconds; this mode, useful in targeting telescope is too slow to comfortably obtain the focus. Also, this camera offers a windowing mode where only a single part of the field is acquired, as in the image above; hence, we obtain a partial image every few seconds.

photosensors. We will see that, for binning mode (as well as for TDI mode, explained in section 2.1.4), it is necessary to distinguish these two notions.

> Binning consists in adding the electric charges contained in several neighboring photo-sensors; all of these photosensors, therefore, are considered to be a single image pixel which received all of the light detected by the photosensors.

Binning could take different geometric shapes: for example, 2×2, which means a grouping of 4 adjacent photosensors to create one pixel: the lines are added two by two and then the resulting pixels from each line themselves are added two by two. We can also find binnings of 3×3, 1×10, etc.

Binning could be done in one of two ways:

- Analog: the electric charges are added in the horizontal register (a binning of several consecutive lines) and in the output stage of the CCD

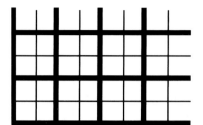

FIGURE 1.27 An extract of a pixel mosaic of a CCD array. The fine lines delineate the array's photosensors. The dark lines delineate the pixels following a 2×2 binning.

(a binning of several consecutive pixels in the same line), then read and digitized just once.

- Digital: each pixel's charge is read and digitized separately and then the computer takes care of adding the values it received.

The second solution is easy to produce and does not need a program modification of the clocks. Then again, the first solution works better, but needs a special function mode for the clocks: several consecutive line transfers must be done before the horizontal register is read, as well as several pixel transfers within the horizontal register with specific cycles before the accumulated charges are read.

One of the advantages of binning is the increase in the signal to noise ratio of each pixel. In the case of analog binning of N pixels, the contents of N pixels are read only once, hence only bringing about one read noise; the signal to noise ratio, therefore, is multiplied by N. In the case of digital binning, each pixel contains the signal and the read noise; during the pixel's addition, the signal (which is a correlated quantity) is added when the read noise (random size) is averaged, resulting in an improvement of the signal to noise ratio of $\sqrt{N}$, which is not as good as in the preceding case.

Binning that gathers $N{\times}N$ photosensors per pixel allows the execution of shorter exposures by a factor of N^2 to obtain the same signal as an image without binning. Another advantage of analog binning is the shortening of the array's read time: for example, the binning of two consecutive lines allows a time gain by a factor of 2 (there being 2 times fewer pixels to read). But digital binning does not allow any time gain since all the pixels are read independently.

The drawback of binning is, of course, the lowering of the array's resolution; in return, the total field is preserved.

We can distinguish three principle areas for the practical application of binning: imaging objects with very faint surface brightness, centering mode, and spectral imaging.

In imaging of objects with faint surface brightness, binning can be used to uncover a very faint object that would otherwise be drowned in read noise.

Binning can equally be used during the telescope's target acquisition phase: during this phase, the image's resolution is not critical, but we hope to obtain images with the highest rate possible, from which benefit on the read time and on signal to noise ratio gain allows the reduction of integration time.

Binning is particularly efficient in the acquisition of spectra: provided that the CCD columns are perpendicular to the wavelength dispersion, a binning larger than 1 pixel on the total height of the spectrum considerably improves the signal to noise ratio and the detection of lines.

TDI mode

TDI stands for time delay integration. This mode consists in letting the image transfer onto the CCD array in the direction of the charge transfer during the read, by synchronizing the transfer speed in the computer with the speed of the image transfer.

An image point begins to generate electrons in the photosensor of the line furthest from the horizontal register; the moment this point moves onto the adjacent line, a line transfer clock cycle is sent so that the integration continues on the same pixel but in the photosensor of the following line and so on until the pixel in the horizontal register is read. The integration time, therefore, is equal to the time it takes the image to pass through the entire array.

This mode has several advantages for astronomy. Firstly, all of the photosensors of one column contribute to the counts of a given pixel. Thus, the faults that can affect a given photosensor (array faults, dust) are averaged over all of the photosensors of the column and are thus practically invisible on the final image.

Another advantage of TDI mode in astronomy is linked to the apparent sky motion: images of the sky undergo a very regular motion due to the Earth's rotation. Instead of motorizing the telescope, which always causes set-up problems and tracking imprecision, it would be simpler to place the camera on a fixed telescope and let the sky go by on the array. TDI mode seems well adapted to automated sky surveys.

This mode of operation supposes that the CCD's columns are perfectly aligned with the direction of the sky's apparent movement and that the telescope's focal length is perfectly known so that the line transfer can be adjusted at the proper frequency.

Finally, note that frame transfer cameras do not have any benefits with TDI mode and those with interline transfer do not permit it: it is therefore the privileged domain of full frame, large format cameras.

Hardware and software sequencers There are two ways to produce clock cycles: by hardware logic or software logic:

- Hardware logic: this signifies that the desired functions are obtained by a combination of logical integrated circuits. The advantage of this solution is that the clock generations are independent of the computer; the camera can therefore easily operate with several different computers. Elsewhere, the computer is free to execute other jobs during the integration time. The electronic module that generates the clocks can be placed near the camera, which limits the length of connection cables. This solution's major inconvenience is its lack of flexibility: it is impossible to modify a cabled circuit to vary its operation; the production of functions such as windowing or binning are a puzzle for circuit designers.

- Software logic: this signifies that the desired functions are produced by a sotfware integrated into a computer; the clocks are program variables transformed into electronic signals by an input–output electronic card interface. In this case, the program must be adapted when the computer is changed, but the required modifications are easily done, as are the windowing and binning functions. Because of its simplicity, almost all CCD cameras destined for amateur astronomers operate according to this architecture.

An intermediate solution is possible: a small microprocessor is placed in the camera's control electronics and is programmed to generate the clocks itself. This solution has the flexibility of the programmed solution while keeping the advantages of the cabled solution. Today, it is not widely used because it is more complex than computer-generated clocks. This solution, however, should have a promising future when the amateur camera market will justify the development of specific complex cards.

1.2.2 *Reading the signal*

Analog processing of the video signal As we have seen in figure 1.25 the video signal is generated in two sequential steps: the first represents a reference level and the second depends on the pixel's electric charge. The electric charge's exact value is the difference in voltage between these two levels. The average voltage of the video signal is about 10 V; hence, its variations, while the pixel's electric charge is at a maximum, are in the order of only 1; the measurement's accuracy we want for this signal is in the order of a dozen microvolts!

To be able to amplify the signal, it is necessary to lower the variations to around 0 V; this is what electronics engineers call removing the DC component. This can be done by subtracting a continuous, perfectly stable voltage, precisely adjusted to the video signal's reference value, or by reading the signal through a capacitor. Both solutions work very well, but each one presents non-negligible fine-tuning adjustment difficulties. Ridded of its DC component, the signal must be amplified in voltage to reach more easily measurable values which are less sensitive to interference. Because of the very faint signals we hope to measure, the amplifier's quality is one of the key factors in a CCD camera's performance.

While it is being amplified, the signal is filtered so as to limit high-frequency noise. One of the camera's important design characteristics is the choice of signal sampling mode. Since the value to be obtained is the difference between the two levels, the easiest solution consists in digitizing, at a given time, the value reached by the first level, and then waiting until the second level reaches its value; both digital values, therefore, are passed on to the computer, which easily does the subtraction. This solution, called *digital double sampling*, is easy to put into operation but has a few inconveniences. Firstly, it assumes two consecutive digitalizations per pixel; unfortunately, the digitizer is, in general, the slowest circuit in the electronic system. The result could be a long read time for each pixel. Also, the digitizing is always blemished by a rounding error, and the act of digitizing twice to produce a subtraction increases this rounding error.

Another solution consists in memorizing the first stage's value in a 'sampler holder' circuit, and while the second stage occurs, the subtraction is made between the value of the second stage and the memorized value of the first in a differential amplifier; all that is left to do is digitize the difference. This solution is called *double analog sampling*. It produces better results than the first, but it uses more circuits and must be done with the utmost care.

The two preceding solutions have a problem: the value of each stage is read only once at a precise moment. To reduce the noise, it would be preferable to read several values per stage and average them; but this would lead to either as many digitizations in the digital difference's solution, which would take an excessive time, or to a very complex circuit in the other case. Instead of taking a large number of measurements, it is possible, for a certain amount of time, to integrate each level's signal into a capacitive circuit. Hence, we accumulate electric charges in one direction during the first level, and in the other direction during the second level. At the end of this double integration the capacitive circuit contains a charge representing the difference between the two levels with effective removal of high-frequency noise. This solution, called *double integration* probably performs best, but it requires that the integration time be exactly identical for each level, which is very difficult to achieve.

Other solutions are possible but are not as common.

Analog and digital signals An analog signal is a variable which can take any value between two limits. For example, an analog voltage with a possible range from 0 to 10 V can take any value between 0 and 10 V. The value of this voltage is the information represented by this variable.

A computer cannot deal with an analog signal. It can only reason with variables that take two distinct values, values that we code traditionally by 0 and 1. An elementary variable is called a *bit*, the abbreviation for 'binary digit'.

Of course, the information contained in an analog variable is much richer than the information contained in a bit, but we can represent it precisely by using several bits. The principle is as follows: Let us take, for example, the coding of an analog voltage between 0 and 10 V. If this voltage is in the upper half of the range (5–10 V), we would decide to encode it as a bit equal to 1, and give a value of 0 in the opposite case (a voltage between 0 and 5 V). The range of uncertainty of the voltage on a single bit is 5 V, or half the interval total. We would now want to know if it is found in the upper half of the field (2.5–5 V or 7.5–10 V) or in the lower half (0–2.5 V or 5–7.5 V); in the first case, we would give a value of 1 to the second coding bit and in the second case, the value would be 0. For example, if both bits used so far are valued at 0 and 1, this means that the encoded voltage is somewhere between 2.5 and 5 V; the uncertainty field is now 2.5 V, or a quarter of the total interval. We can continue the encoding in this manner: each time a supplemental bit is added, the range of uncertainty is reduced by a factor of 2. A bit allows us to encode 2 different values: 2 bits, 4 values; 3 bits, 8 values; 4 bits, 16 values; etc . . .

Computers generally handle groups of 8 bits called *bytes* which allow the representation of $2^8 = 256$ different values.

Table 1.3 gives the number of values represented in relation to the number of bits used. We can ascertain that precision increases very quickly with the number of bits.

Digitization of the video signal

After we have removed its continuous component, filtered it, and amplified and sampled it, the video signal is still in an analog format, incapable of being processed by a computer. It must therefore be converted into digital information. This is the role of the analog–digital converter (ADC).

The ADC is one of the most expensive electronic circuits of the CCD camera. Currently, 8, 10, 12, 14, and 16 bit ADCs are available (which means that the digital information output is coded on 8, 10, 12, 14, or 16 bits).

Table 1.3

Number of bits	1	2	3	4	5	6	7	8
Number of represented values	2	4	8	16	32	64	128	256

Number of bits	9	10	11	12	13	14	15	16
Number of represented values	512	1024	2048	4096	8192	16384	32768	65536

Unfortunately, the conversion times are, in general, longer as the number of bits increases. More precisely, the price of the converter increases (very quickly) with the number of bits and with its speed. But progress in this field is relatively quick and we are just now (1997) beginning to find 16-bit converters working in less than 10 µs, costing less than £140/$200, which offers a very interesting solution for CCD cameras.

A converter's number of bits is generally declared by the CCD camera manufacturers: 8 bits for the ST4, 12 bits for the LYNXX, 16 bits for the ST6 and HPC-1. It is a very important criterion to take into account when judging a camera. We will return to this in our chapter on CCD camera performances.

1.2.3 *The electronic design of a CCD camera*

Several types of architecture are possible in the making of a CCD camera. Although we will not present an extensive report, we will, however, present a few of the most recent solutions.

We will look exclusively at cameras whose clocks are generated by computer software, which is the easiest solution to produce as well as being the most popular among amateur astronomers.

It is necessary to place a circuit board in the computer, containing an output circuit that converts the program's instructions into electric variables. Binary electric variables obtained by standard circuits provide a voltage of 0 V for the 0 level and 5 V for level 1 (these values are characteristic of TTL technology used by computers). It is necessary to transform them into signals whose values correspond to the required levels of the CCD's clocks (see the examples given in section 1.2.1). Not only must the voltages be respected, but the clock signals must have an adequate drive capability and transmission time which is neither too long nor too short between the high and low levels. Therefore, we use 'clock timing' circuits to transform the TTL signals coming from the computer into signals that are usable by the CCD; these circuits are the subject of a new electronic card, directly connected to the CCD's input.

The continuous component of the video signal must be subtracted and then amplified and filtered by analog electronic circuits; then it is sampled and digitized. Its digital value, therefore, is represented by a specific number of TTL standard bits (0–5 V) accessible to the computer through an input circuit that can be implemented on the same card that has the clock's output circuits. Nowadays, one can even find input–output circuits that produce both of these functions in a single component.

Where are these circuits placed? We often try to put as few circuits as possible in the camera head, which contains the detector, firstly, to reduce its mass and volume (it will be fixed onto the telescope), and secondly to avoid creating heat sources in terms of the CCD (all electronic circuits dissipate heat and we will see in section 1.3 that the CCD must be cooled). However, the clock timing circuits must be placed as close as possible to the CCD so that the signals provided are not altered by too long a connection (weakening of signal, appearance of interference). For its part, the video signal in its analog form is very sensitive to interference: it must be amplified and digitized as close as possible to the CCD. The electronic engineer's desire to place the circuits as close as possible to the CCD goes against the user's worry of having a lightweight camera body and a computer well away from the telescope (in a building situated 20 meters from the dome, well heated, . . .).

Figure 1.28 represents a possible compromise: the card inserted into the computer only contains input–output circuits; the printed circuit placed in the camera head only supports the CCD, the filtering circuit of the continuous component, and a pre-amplifier. An electronics casing, placed a few dozen centimeters away from the camera body, contains a clock timing and a video signal treatment card (amplifying, filtering, sampling, and conversion). The computer, therefore, can be located at some distance from the electronics casing, but a large number of cables would have to be connected between them. Several variants are possible: in the goal of simplification, and hence reduced costs, amateurs have built cameras that 'talk' with the computer through the printer's connection port. We therefore save an input–output card, but the information transfer is relatively slower. In the same vein, the ST4 and ST6 cameras made by SBIG in the USA are connected to the computer's serial port; this system is also more economical but slow (22 seconds are needed to read the image from an ST6!).

Still with the goal of manufacturing at low costs, SpectraSource Instruments has adopted another solution for its LYNXX cameras: the electronics module is

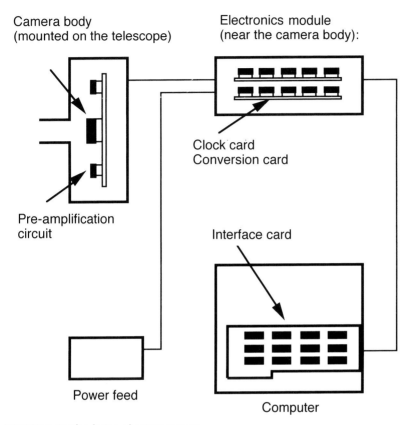

Camera body
(mounted on the telescope)

Electronics module
(near the camera body):

Clock card
Conversion card

Pre-amplification
circuit

Interface card

Power feed

Computer

FIGURE 1.28 The design of a CCD camera.

eliminated and the analog–digital converter is placed on the computer's interface card. A shielded cable brings the video signal in analog form directly to the
camera body until it reaches the interface card; this cable also carries the clock
signals, the shutter command, and even the Peltier module's current (see
section 1.3). This solution, which is an electronics engineer's nightmare, actually works perfectly; its only drawback is the limitation of the cable's length
between the camera body and the computer: 15 meters for some cameras, 7.5
meters for others.

Other cameras, on the other hand, like the HPC-1, have the essential
electronics implemented in the camera body so as to give preference to the
signal quality over compactness and cooling. The Cyrocam cameras obey
the same principle, but in addition have a cabled sequencer also implemented in the camera body, which makes them particularly heavy and
bulky.

1.3 Thermal and mechanical aspects

1.3.1 *General points concerning cooling*

The need for cooling We have seen in section 1.1.5 that unwanted electric charges are generated in the photosensors independent of incident light. Even an MPP technology CCD produces, at ambient temperature, a few thousand electrons per pixel per second. Saturation appears in a few minutes (a few seconds if the CCD is not MPP). Also, a non-saturated image, for example, that contains 100 000 electrons per pixel of dark current would represent a thermal noise of 300 electrons (the square root of the dark current); therefore, if we know how to remove the average dark current value, we do not know the residual difference, which is thermal noise, of the signal to be analyzed; 300 electrons for the only thermal noise in an exposure of a few minutes is often inadmissible in astronomy.

Fortunately, dark current generation depends strongly on the temperature: it is divided by 2 each time the temperature is lowered by about 6 °C. If the CCD is cooled by 60 °C in relation to the ambient temperature, the dark current, therefore, is 1000 times weaker. For this reason, almost all cameras used in astronomy are cooled; the few that are not cooled are very inexpensive and are not suitable for detecting faint objects.

The exposure times commonly used in astronomy, and the required detection levels mean that the CCD be cooled so as to limit its dark current.

It is not always necessary to cool to very low temperatures. Everything depends on the thermal current's level that can be tolerated and the CCD's technology. An MPP detector gives excellent results at −40 °C. The most sophisticated cameras, used by professional astronomers, are generally cooled around −80 or −100 °C. Beyond that, any gain in dark current is weak and we run the risk of actually diminishing the CCD's sensitivity. Also, we have to avoid too low a temperature, and especially a drastic temperature variation, which could damage the detector; anyway, the manufacturers rarely specify the low-temperature performance of their CCD. Finally, the lower the desired temperature, the more the camera body's insulation and cooling system become complex and the more the camera loses sensitivity in the red.

The CCDs used in mass market video cameras are not confronted with this problem: each image is exposed for 1/25 s, which limits the number of thermal electrons generated. And since the scenes filmed provide a lot of light, the thermal noise is negligible compared to the photonic signal. It is not necessary, therefore, to cool these CCDs.

The production of cold

In order to cool the CCD detector, several solutions are possible: liquid nitrogen, carbon
dioxide slush, and Peltier effect coolers.

An efficient method consists of equipping the camera head with a liquid-
nitrogen-filled reservoir. A piece of copper (a very good thermal conductor) is
placed behind the CCD and crosses the reservoir tank wall thereby soaking it
in liquid nitrogen and bringing the cold to the CCD. This process is so efficient
that it is often necessary to heat the CCD by heating resistors so its temperature does not decrease too much. Liquid nitrogen is at −196 °C. These resistors, which are easily controlled, have the principle role of regulating the
detector's temperature around −100 °C. Liquid nitrogen is widely used by
professional astronomers because of its efficiency. It is, however, impractical
to use because of the necessity of an expensive cryostat, which is also cumbersome and heavy. Also problematic is the difficulty an amateur has in obtaining
liquid nitrogen, storing it, and pouring it into the cryostat. Because of these
difficulties, nitrogen cooling is not the best solution for amateurs.

The same principle can be used when replacing liquid nitrogen with dry ice.
The reservoir must be equipped with a spring-loaded mobile partition wall so as
to compact the dry ice against the cooling element as it evaporates. The temperatures reached are not as low (the temperature of dry ice is −80 °C), but still sufficient for our needs. Carbon dioxide snow is quite easy to make from a cylinder
of carbon dioxide gas which is evacuated through a small nozzle. This is the principle of some fire extinguishers. It can also be purchased in the form of blocks or
bars which can be kept for a few days in a freezer. Unfortunately, the cryostat is
still complex, heavy, and cumbersome, and the frequent refilling quickly
becomes annoying. For this reason, dry ice is not used much by amateur
astronomers. The most flexible solution is achieved by thermoelectric coolers.
These coolers are called Peltier effect modules, named after their creator, or now
simply Peltier modules. They are a type of heat pump controlled by an electric
current. Their efficiency is limited, but their ease of use and their reduced bulk
make them the best choice for the amateur astronomer (section 1.3.2 is devoted
to their description).

The thermal load The cooling system's performance is defined by the temperature the CCD must reach and by the thermal power that must be dissipated; the
latter includes not only the thermal loads generated by the electronic circuits
but also the heat produced by the entire CCD system, which is much hotter
than the detector. It is essential to limit these thermal leaks in order to limit the
amount of thermal power to be removed.

We will examine the different sources of thermal loads:

- The heat loss of electronic components. To work, each circuit consumes a specific current, which produces a thermal loss by the Joule effect (which is simply the physical effect described by the familiar high-school formula: the dissipated heat is the product of the square of the absorbed electric current by the internal resistance of the circuit). Of course, the camera body contains the CCD itself, which constitutes the prime heat source. Electronics engineers would also like to place the clock's controls, the video signal amplifier, the analog signal processing system, and the converter closer to the CCD, ideally in the camera body. Overburdening the camera body is this solution's limitation, but the major objection is the production of a prohibitive thermal load by all its circuits. Each one of them dissipates a few hundred milliwatts, so only a bare minimum is left around the CCD, in general, a single signal pre-amplification stage.

- The conduction between the CCD and the camera body casing, which is at room temperature. To limit this conduction, the CCD is equipped with insulation materials. But an unavoidable conduction source is furnished by the electric wires that ensure a link between the CCD and the camera body's output pins: the metals are electric conductors, but also very good thermal conductors. The conduction exchanges between these wires can be reduced by the use of very fine wires that are as long as possible; then again, a heat engineer's concerns are exactly opposed to those of an electronics engineer.

- The convection of the gases enclosed in the camera body, between the cold source, which is the CCD, and the hot source, which is composed of the body's walls. As in conduction, the transferred thermal power is proportional to the temperature difference between the CCD and the wall. An excellent way to eliminate transfer by convection is to place the inside of the camera body under vacuum. But this solution demands excellent seals for the camera body's walls, which is sometimes difficult to achieve at the points where the wires cross the walls. Without a vacuum, one should try to limit the cold surface area.

- Radiation. Physics teaches us that all bodies radiate an energy proportional to their surface area and to the fourth power of their temperature (Stefan's law). The CCD thus receives radiation (in the infrared, around $10\mu m$) from the tank's walls, but itself re-emits little energy since it is cold. The balance of this exchange shows that the CCD tends to heat up through radiation; to limit this effect, one should try to minimize the surface area that needs to be cooled.

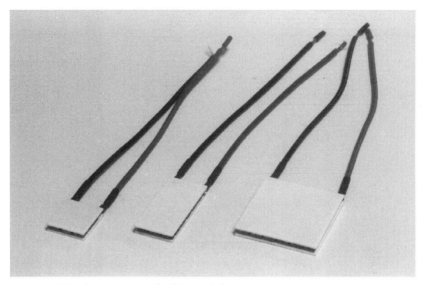

FIGURE 1.29 A few examples of Peltier modules.

1.3.2 *Cooling with thermoelectric elements*

Peltier modules A Peltier module looks like a flat element, a few millimeters thick, whose two main sides, either square or rectangular, are made of ceramic (see figure 1.29). Two wires enable the supply of an electric current. When an electric current is sent to the module, heat is transferred from one of the sides (cold) to the other (hot).

Unfortunately, this transfer is not very efficient: the electric power consumed is greater than the thermal power being pumped. The thermal power on the hot side is the sum of the power pumped from the cold side and the electric power consumed and dissipated in the form of heat. We must therefore remove from the hot side heat greater than that absorbed by the cold side.

Peltier module characteristics A Peltier module is characterized by the maximum temperature difference it allows between its cold side and hot side, and by the maximum thermal power it can pump. This information is usually supplied by the manufacturers. Unfortunately, these two types of performance are incompatible: the maximum power pumped is reached when the temperature difference is zero (which is not the desired goal!), and the maximum temperature difference is reached when the pumped power is zero (which never happens!).

Figure 1.30 shows the characteristic curve of a Peltier module. A point should be chosen that will allow us to pump the necessary power while

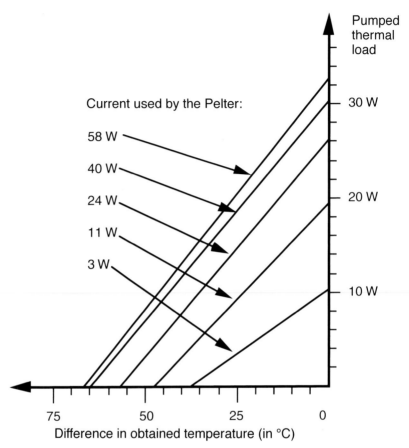

FIGURE 1.30 One difference in temperatures obtained by a Peltier module (CP1,4-127-05L model from Melcor) in relation to the pumped thermal load and the current strength.

achieving a reasonable degree of cooling. We can increase both of these parameters by increasing the current passing through the Peltier module, but this is detrimental to its efficiency. To optimize the efficiency, the magnitude of the current is generally chosen between 40% and 60% of the module's maximum capacity. We can also choose weaker Peltier modules. Whatever the designer's choice, we can see that in the Peltier solution, the thermal power to be evacuated is a fundamental parameter, which explains the efforts made elsewhere to limit the thermal load (see section 1.3.1).

The temperature attained with regard to the CCD is equal to the temperature of the hot side minus the Peltier system's temperature difference. To obtain the lowest temperature, therefore, we must:

- find the system that gives the largest temperature difference possible, and

- ensure that the hot side's temperature is not too high in relation to the ambient temperature; we must therefore limit the power to dissipate (which goes against the goal of optimizing the temperature drop) and choose a higher-performance cooling system.

To increase the temperature difference between the CCD and the hot side, it is possible to stack several layers of Peltier modules in a series: the final temperature differential between the CCD and the camera body's hot side is the sum of the temperature difference of each Peltier module. But this solution, if it is widely applied to CCD cameras, is not the miracle solution because of the Peltier's poor efficiency. In fact, the coldest Peltier, in contact with the CCD, absorbs the thermal power from the CCD, which is weak, and can then provide an interesting temperature difference; meanwhile, the hot side removes more significant power and is thermally connected to the cold side of the next Peltier module. The latter must absorb a greater power, therefore providing a weak temperature difference or in turn dissipating even more heat. It can be seen that a large number of layers, although it improves the cooling effect, it also generates a lot of heat that has to be dissipated.

The optimization of a series of Peltiers is a delicate problem: their different elements and operating points must be astutely chosen in order not to get a catastrophic efficiency. Figure 1.31 shows the result of optimizing multilayer systems. The calculations were carried out for an initial power of 2 W. We assume that:

- The number of levels must be chosen in relation to the desired temperature difference (ΔT).

- The price for requiring a high temperature difference would be having a considerable amount of heat to remove.

We can see that the lowest temperature when it comes to the CCD is not always supplied by the system that gives the largest temperature difference: in fact, if the system dissipates too large a power with respect to the performance of the hot side's radiator, the hot side's temperature can rise to a value higher than the temperature difference gain.

The removal of heat The back of the camera body, to which the Peltier's hot side is fixed, receives a significant amount of thermal power: it is necessary to equip it with a heat exchanger system in order to remove the heat. This heat removal system is fundamental; if it is inadequate the hot side's temperature will rise, which as a consequence will raise the CCD temperature, since the Peltiers provide a constant temperature difference between the two. A heat

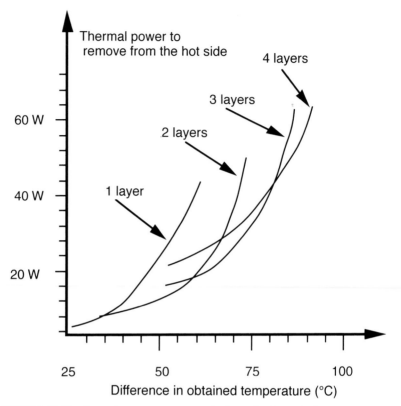

FIGURE 1.31 The thermal power to evacuate the hot side of a multi-layer Peltier system optimized in relation to the needed temperature difference, for a nominal charge of 2 watts.

exchanger system removes thermal power proportional to the temperature difference between its walls and the cooling fluid (air, water, . . .); it is therefore characterized by the number of degrees (the difference between these temperatures) required to evacuate 1 watt. This performance is comparable to that of stacked Peltier modules: if the radiator needs 2 degrees to evacuate 1 watt, a stronger Peltier power leading to a temperature difference better than 5 °C but at a dissipation higher than 4 W leads to a rise in the CCD's temperature of 3 °C!

There are three possible solutions to remove heat from the hot side:

- The hot side is equipped with a finned radiator and heat is dissipated into the air by natural convection. This solution is the simplest method and is widely used in commercial cameras intended for amateur astronomers. Unfortunately, it is not very efficient: 0.6–5 °C/W according to the size

FIGURE 1.32: this LYNXX camera, seen from the back, shows that the Peltier module's hot side calories are evacuated with the help of a finned radiator through a simple convection effect with exterior air.

and shape of the radiator; but it is not easy to place a gigantic radiator on a camera body.

- The hot side is equipped with a finned radiator, but with a small fan blowing on the radiator. This solution, called forced convection, is ten times more efficient than the previous method. The problem with it is linked to the presence of the fan, which is bulky but more importantly generates vibrations. Forced ventilation is the ideal compromise between efficiency and simplicity if the camera is mounted on a large and very stable telescope; it should not be used on a small unstable telescope. One could always place the fan on the ground and direct the blown air toward the radiator through a flexible hose, but this is a source of complications.

- A water (or coolant) circulation radiator is placed on the hot side. This is very efficient: 50 W evacuated for a 1 °C temperature difference between the hot side and the water. But this solution requires the use of a water container, a pump, and two circulation pipes between the camera body and the water tank, and the possible piping of water from an auxiliary radiator to cool the water which is already warm. The system could become formidable (in terms of efficiency and constraints) if one thinks of filling the water circulation container with ice cubes in order to keep it near a temperature of 0 °C. We will skip pump starting problems, dome flooding at the smallest blunder, frozen tubes during winter observations . . . the efficiency must be paid for somewhere.

The thermoelectric elements (Peltier modules) are a flexible and efficient way to obtain a reduction of 30–60 °C in the CCD's temperature. But they need a secondary cooler (finned or water-cooled radiator) to evacuate the heat linked to their functioning.

The power supply and temperature regulation of Peltier modules The power supply for Peltier modules is current regulated since their operating mode is determined by the amount of current passing through the modules. When the power is turned on, we try, in general, to progressively increase the current's magnitude to avoid giving the CCD detector a thermal shock. Moreover, the Peltiers' performance and life span depend on it.

It is easy to attach a thermocouple to the CCD detector to measure its temperature (the Thomson 7895 CCDs even have an integrated temperature detector!), and regulate the Peltier's current intensity to keep the CCD's temperature at a constant value. Here, there are two conflicting schools of thought. Some believe that it is essential to control the CCD's temperature in the most precise way possible. For example, the OMA4 camera proudly advertises regulation to the $1/100$ °C (it should be said that this camera represents, on all points, exceptional performance but also a price tag that places it out of reach for all amateur astronomers). The motive is that the dark current depends strongly on the temperature; the detector, therefore, must be at the same temperature when the sky's image is taken as when the dark image is taken, which will be used to remove the dark current from the former.

As far as we are concerned, we think the exact opposite: the preceding argument is acceptable if the image treatment software can only remove one image (the one containing the measured dark current) from another (the sky one). But the software now available to analyze images use functions that know how to adapt the dark current's level adopted in a thermal (or dark) image before removing from it an image of the sky produced in completely different circumstances. With this software, a 'card of darks' can be produced for the camera with very long exposure times, and by averaging several measurements it can be used for all the following images (see chapter V section 2.2.3). In this way the treatment quality is optimized and it is no longer necessary to produce dark images every night for each exposure time. Therefore, it is not necessary to regulate temperature. But this raises some difficulties: if the regulation is done by switching the modules on and off rather than by adjusting them, it can generate interference during the CCD reading, and therefore if we want to regulate the temperature, the Peltier modules cannot operate at their maximum performance, thus the CCD loses several degrees of temperature. Consequently, we are in favor of a non-regulated cooling system.

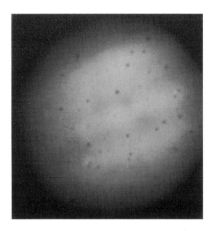

FIGURE 1.33 This image of Jupiter is strewn with tiny black dots characteristic of frost specks on the CCD window.

1.3.3 *The problem of frost*

Everyone knows that when one removes a bottle from the refrigerator it is quickly covered by dew from the ambient air, which is warmer and more humid than the bottle, which is only a few degrees colder. We imagine that this happens to the CCD if it is left in ambient air as its temperature is lowered by several dozen degrees below zero. It is not a layer of dew but a layer of frost that would instantly cover the CCD, thereby preventing any light from reaching the detector.

To avoid frost, it is absolutely necessary that the CCD be placed in an area devoid of water vapour. Two classic solutions exist: either the CCD is placed in a vacuum, or it is placed in a dry nitrogen atmosphere (or dehydrated air).

Both of these solutions require that the camera body be airtight, to avoid any ambient air penetrating the interior. But a good vacuum seal is difficult to produce; if the use of screw seal joints allows the realization of a good air seal of the gap between the cover and the body of the camera, the passage of the electrical connections raises a problem which is rarely solved in a satisfactory way, with the exception of using special airtight connectors whose prices are often a deterrent to their use.

The dry nitrogen solution appears to be the most economical: the filling of the camera body with nitrogen enables the body to have an interior pressure slightly higher than the outside atmospheric temperature. In these conditions, an acceptable air seal can be achieved without using the special connectors. But we are not safe from nasty surprises if the camera is used at high altitudes, such as the Pic du Midi Observatory, for example: the atmospheric pressure at 3000 m is only 2/3 of the pressure at sea level (and hence of the pressure of the nitrogen inside the camera body); in a few days, residual leaks are sufficient to allow the necessary quantity of nitrogen to escape so as to balance the pressures on the inside and outside. Once the camera is returned to a low altitude, the inverse phenomenon occurs: the camera's interior, which established itself at a lower pressure, slowly returns to equilibrium by taking in ambient air, which, itself, is filled with humidity.

Cameras filled with nitrogen are:

- either purged then filled with nitrogen, after having been sealed, thanks to a valve; this system is very simple but requires finding a valve that is not too expensive but very airtight; or
- closed by the manufacturers in a nitrogen atmosphere, which requires the use of an airtight 'glove box', in which we can replace air by nitrogen or argon to allow the placement and manipulation of the camera head.

Generally, this equipment is not easily accessible to amateur astronomers. In the case of a camera leak, it should normally be returned to the manufacturer for a new filling of nitrogen. Of course, it is strongly recommended never to open a camera body that has been filled with nitrogen.

Putting a body under vacuum has a great advantage: it eliminates both the risk of frost and heat transfer by convection at the same time. The major problem is in obtaining a seal that is now efficient not between two gases at the same pressure as before, but between two environments which have a difference in pressure of 1 bar. There are two ways to overcome this problem:

- either we hope that our camera will keep its vacuum for years, which is technically possible, but requires a very expensive mechanical construction; or
- we admit that there will be leaks, and we permanently suck the air from the camera body with a vacuum pump.

It is desirable to obtain a vacuum in the order of 10^{-1} to 10^{-2} torr (1 torr is equal to a pressure of 1 mm of mercury). The necessary vacuum pump, therefore, costs between £300 and £600. However, this solution seems more economical and more certain than a permanent vacuum. An intermediate solution, which would consist in keeping the vacuum airtight for a few weeks, does not seem satisfactory: because the amateur astronomer will not be sending their camera back to the manufacturer every few weeks or every month to put it under vacuum again, they would have to equip themselves with a pump. Once bought, it might as well be used at the time of observation.

If the camera works under a vacuum, one would aim to pump it before cooling the CCD, and to let it heat up before stopping the vacuum pump. Without these precautions, the detector can be temporarily covered with frost, which could leave traces.

It is inadvisable to take a commercial camera that is meant to be filled with nitrogen and to put it under vacuum without consulting the manufacturer: indeed, some electronic circuits within the body are no longer cooled by convection and can also produce corrosive gases, which could lead to some unwanted surprises.

Finally, note a particularly robust solution used in the ST6 camera by SBIG: the camera body is neither filled with nitrogen nor under vacuum but contains ambient air (which simplifies the problem of airtightness); small packets of silica gel are placed beside the detector and are supposed to protect it from humidity, which is efficient as long as the silica gel does not saturate. Once the CCD frosts, the user must open the camera and change the silica gel packets.

1.3.3 *The camera body*

We have just seen the different functions (electronic, cooling, airtightness) the camera body must fulfill. We have examined the different possible solutions, while bringing to mind the efficiency objectives and cost constraints, bulkiness, weight, and user-friendliness.

As in the electronic design of the camera, several solutions are possible for the body. Figure 1.34 shows the typical layout of a camera body. The reader should recognize the different components already mentioned, and could identify them in this structure with the cameras they might have to use.

1.4 The computer environment

1.4.1 *The computer*

The computer is an integral part of the CCD system. During the image acquisition, it is used to send the commands for reading, reception, visualizing, and storing the digital images. Once the images are stored in the computer, they will be processed in order to extract the best part. The computer's presence, therefore, is unavoidable since it is an interface between the user and the CCD camera. Thus, one must equip oneself with a computer that is powerful enough, and user-friendly but not too expensive!

In terms of the quality to price ratio, PC-compatible computers are the best products. They are the most widely used and every CCD camera on the amateur market can be operated by them: we will only deal with PC compatibles. This

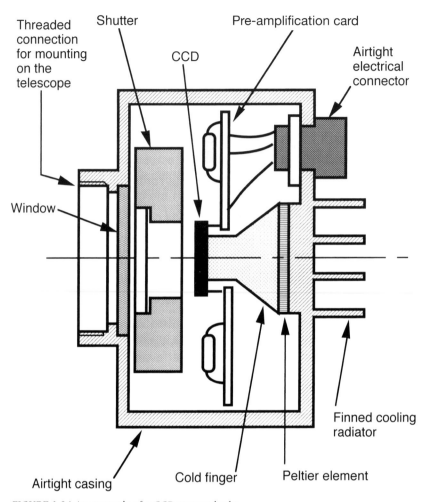

FIGURE 1.34 An example of a CCD camera body.

family of computers should be put into context. At the beginning of the 1980s, IBM introduced onto the market PC-compatible personal computers based on the Intel 8086 microprocessor. A few years later, this model was replaced with the 80286 and then by the 80386. The 1990s have seen the appearance of the 80486 and, finally, the Pentium. The purchase of one of these computers must be accompanied by a series of indispensable options to produce and process CCD images.

One must keep track of a few supplementary constraints in order for the computer to control a CCD camera. In particular, some CCD cameras operate with an interface card inserted into the expansion slots inside the computer. One must also, therefore, verify that these slots exist and ensure that they have enough connections to accept the card. There exist 8, 16, and 32 bit slots. Get information from the camera dealer.

Understandably, portable computers do not always have available space inside. However, if you can use a portable, which is preferable when the computer has to be moved each observation session, verify that it is equipped with a screen that correctly displays gray levels (preferably a color LCD array screen) and has a long-life battery.

Some manufacturers, whose cameras do not need an interface card, make the possible use of a portable ('lap top') a sales argument. Even if a portable computer is attractive, avoid falling for the purely commercial sales trap, because this solution does not only have advantages. Indeed:

- Using a portable generally requires a dialog through the parallel port or the serial port, which are each slower than the interface card; the conceived camera, therefore, has a slower read, which is penalizing when there is a large amount of work.

- Portables are actually very often limited to the VGA standard, whereas larger arrays require Super VGA software.

- It is still possible to use a desktop computer, well suited to guiding a camera equipped with an interface card, in the countryside using a car battery as a power supply. This solution is efficient and not very costly. Certainly the desktop computer is heavier and more cumbersome than a portable, but it is surely less so than the telescope which is obliged to be part of the trip!

Given the equipment available on the market (in 1997), we can assume that a computer capable of processing digital images must be equipped, at the minimum, with a 166 megahertz Pentium microprocessor, 32 megabytes of RAM, a 1.2 gigabyte hard disk, and a 1 megabyte Super VGA graphics card (with a 1024×768 resolution with 256 colors) to the PCI standard. The color monitor must be a multisynchronous analog type (in non-interlaced mode for the 1024×768 resolution) and equipped with horizontal and vertical adjustment features.

The golden rule before purchasing such a computer is: first choose the CCD camera and the image processing software that will be used. Then, decide on the minimum configuration of the computer you must buy and get the technical specifications of the camera and a demonstration version of the software. Finally, try the software demos on the computer you have chosen to buy.

1.4.2 *Software*

PC-compatible computers are all equipped with an operating system, usually DOS (Disk Operating System). It is also possible that you have Windows 3, which is management application software made user friendly by its use of windows and menus. Since 1995, many computers have been installed with the Windows 95 system. Some software uses only DOS, while others require Windows. The newer software is generally designed to be used with Windows 95 and will not be executable with DOS. Become well informed before launching the program. Also, as good as the acquisition and image treatment software are, they generate a lot of files on the hard disk. Therefore, it is very helpful to learn to use file management software.

Acquisition software Each CCD camera model has its own acquisition software. Its goal is to allow the user to initiate the CCD's exposure commands, capture images, and store them onto the computer's hard disk. Nevertheless, it is always necessary to match image processing software with the acquisition software.

The standard acquisition of an image consists of integrating the light for a given amount of time, then transferring all of the CCD array's pixels to the computer's memory. Then the image is displayed on the monitor, and, if it is satisfactory, saved to the hard drive. It is preferable that images can be saved in the standard FITS format (Flexible Image Transport System). Several variants of acquisition modes may be needed.

We have just described the single exposure mode: the operator gives the CCD read command for each exposure. During the targeting of an object in the field, it is useful to have an automatic mode. It is sometimes useful to use only part of the CCD's surface so as to save space on the hard disk or to read the array quickly. In this case, the software must allow windowing. Acquisition in binning is another function that consists in regrouping several adjacent photosensors into a single one. There are several other modes (TDI, multiwindowing with or without mobile windows, photometric or rapid astrometric observations, and so on).

The image acquisition software must be able to display the images with at least 16 levels of gray in order to detect the smallest of problems (interference, frost, etc.) so that they can be resolved immediately. It is necessary to be able to display and move a reticule over the image to measure the number of ADUs (defined in section 4.12) and have a statistical function that searches for the maximum value of the image in order to detect if one of the pixels has saturated during the exposure. It is important to arrange a few simple image analysis tools: preprocessing and an unsharp mask (described in section 5.5.1) are very useful options for quickly verifying the images' quality. Finally, all acquisition software must be user friendly and easy to use. The possibility of having access to the functions through menus and dialog boxes is often preferable, on condition that the text is clear.

Image processing and display software Acquisition software allows the capture of raw images. Image processing software must therefore be able to read the images stored on disk by the acquisition software. Hence, we must always insist that the image processing software be able to read FITS format images.

The raw images are marred by several blemishes that must be removed. For example, because of vignetting, it is impossible to make photometric measurements on raw images. The processing software, therefore, must be able to correct the raw image's faults; this is the preprocessing operation (described in section 5.2). The preprocessed images can then undergo an infinite number of treatments, among which we can distinguish two large families:

- The first type of processing consists in transforming the image's pixels to create a new image on which the useful details will be amplified. This is truly a full image processing operation. In this category are included point to point operations, geometric transformations, convolutions, and image restorations.

- The second type of processing consists in extracting some digital values from the image; this is the operation of data reduction. For example, on a square image of 100×100 pixels, there are $10\,000$ bits of elementary information. If we are looking to measure the magnitude of a single star in a field, we would reduce the data. In this category are included photometric, morphological and astrometric analysis functions.

All processing software must be capable of displaying the images with the best graphics modes possible. For a monochrome display, insist on at least a 64-level grayscale. In addition to the grayscale, palettes of false colors are sometimes useful in certain cases. However, one must not be overly impressed by software that boasts an astronomical quantity of false color palettes. A strict test simply consists in displaying some low-contrast images using the gray palette; this will reveal a software's display faults. As for the displaying of images in three-color mode (color imagery), always ask to see a demonstration in order to judge the quality of the color rendering since several algorithms exist for restoring the entire image with a restricted color palette.

All image processing software works according to simple and elementary functions. For example, to display a raw image saved onto the hard disk, we must call on two elementary functions. The first function consists in loading the image into the computer's memory and the second in displaying the image in memory. During the loading stage, the software will ask you to indicate the file name that contains the image. During the display stage, it will ask you to indicate the visualization thresholds.

Table 1.4 *The key words LOAD and VISU are elementary functions that serve, respectively, to load an image into memory and to display it. These functions imply a dialog with the software through the intermediary of a question and answer mode*

Elementary function	Logic question
LOAD	Name of image file?
VISU	Lower limit? Upper limit?

There are two large families of image processing software: those that execute elementary functions from drop-down menus and those that ask the user to enter, by keyboard, the key words corresponding to the desired elementary function. The key word method needs a good knowledge of the keyboard and knowledge of the key words. This often discourages the beginner who obviously feels more comfortable being guided by menus with explicit titles. The better image processing software allows the use of either menus or key words.

Working with key words is not always inconvenient. Firstly, let us stress that it quickly becomes laborious, almost painful, to permanently be clicking on menus once we begin to master the sequence of elementary functions that need to be applied to produce some lengthy processing. Secondly, note that several elementary function sequences must be repeated over the course of one processing session. The best software, therefore, allows us to program the function sequence in order to create, at will, real macrofunctions. In this case, the macro functions (which can even become little programs in certain cases) are written from the elementary functions' key words.

In conclusion we recommend that those who are not familiar with computers equip themselves with menu-based software. On the other hand, if you are currently using computers, it is preferable to choose software that also functions by key words, since you will quickly be able to program the macrofunctions that will allow you to gain some precious time. For instance, doing the preprocessing by drop-down menus can take several minutes, while the programming of a macro function will allow the pre treatment to be done in less than 30 seconds. Imagine the time saved when you have hundreds of images to process!

2 CCD camera characteristics and performance

2.1 Geometric characteristics

As we have seen in the previous chapter, the CCD chips used in astronomy are black and white arrays. We will only deal with these in this text.

2.1.1 *The number of pixels*

One of the most important characteristics of the CCD camera is the number of pixels in an array. It is this number that dictates the quality and richness of the spatial information of the image: it is the link between the area covered by the CCD and its resolution; this concept is sometimes called 'spatial dynamics'. It is possible to adapt the area covered or resolution by choosing optics of an appropriate focal length. However, the area covered and resolution are infinitely linked by the number of pixels.

For example, if we were to use a focal length such that one pixel corresponds to 2 arc seconds, which is a reasonable choice for images of faint objects, a Thomson 7852 array (208×144 pixels) covers 6.9'×4.8', a Thomson 7863 array (384×288 pixels) covers 12.8'×9.6', a Thomson 7895 array (512×512 pixels) covers 17.1'×17.1', and a Thomson 7896 array (1024×1024 pixels) covers 34.1'×34.1'. It is obvious that a larger array is more desirable. Hence, CCD prices tend to increase rapidly with the array's size. The same applies to cameras. The other difficulty about a large array is the time it takes to read. For the same given data, the read time is directly proportional to the total number of pixels; therefore, the read time for a Thomson 7896 array would be 35 times longer than for a Thomson 7852 array. One should not be too hasty in dismissing the smaller arrays, though: with optics having a focal length of 1800 mm, a LYNXX camera (192×165 pixels) covers an area of 5'×5' with a 1.7″ resolution. This permits one to produce quality images for nearly all the faint objects: reading through the NGC catalogue confirms that a small minority of these objects have a size greater than 5'. In fact, mass market publications and years of practice in astrophotograpy have familiarized amateur astronomers with spectacular objects that are easy to photograph: M8, M13, M31, M42, etc; these are quite large and bright. Smaller and fainter objects, although more

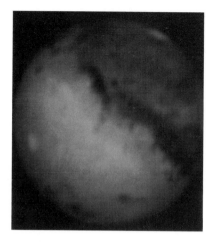

FIGURE 2.1 An image of Mars during its opposition of September 1988 taken with a 1 meter telescope from the Pic du Midi Observatory with a Thomson 7852 array. Notice the presence of a white cloud cap covering the Olympus Mons Crater. Reference: Jean Lecacheux.

numerous, are, unfortunately, unrecognized, since they are much more difficult to photograph. Now, the sensitivity of CCD cameras makes these objects easily accessible to amateur astronomers.

Modern software now makes it possible to manipulate an image, for example juxtaposing several images together to produce mosaics. These methods also allow images of objects that are larger than the camera's field of view. Of course, the time dedicated to observation will be multiplied by the number of images necessary; it will be about the same amount of time as needed for image processing.

Planetary imaging is a field in which a small array can give excellent results: as of early 1994, the best images in the world of Mars, Jupiter, and Saturn (from Earth) were obtained by Jean Lecacheux's team working with the 1 meter telescope from the Pic du Midi Observatory with a Thomson 7852 array, one of the smallest used in astronomy. A field of view of only 20 seconds of arc is sufficient to cover the entire planet Mars and permit sampling to a resolution of 7 to 8 pixels per arcsecond, even with a small array, which is already very good for high resolution.

Photometry is another example where the area covered is not an important parameter. Of course, it is much more agreeable to work with an extensive field in order to easily identify a specific star and obtain, if possible, reference stars on the same image. Before the appearance of CCDs, all photometry was undertaken using photodiode photometers and photomultipliers, which can only measure one star at a time.

On the other hand, astrometry needs good spatial dynamics. In fact, it is necessary to have both a high resolution and a large field of view in order to include reference stars in the image. Small arrays, in general, are inadequate, but the

FIGURE 2.2 A mosaic image of M42 composed of four images obtained with a LYNXX camera and a focal length of 500 mm. Photos by Patrick Martinez and Pierre-Michel Bergé; image processing by Cyril Cavadore.

situation could evolve with the availability of faint star catalogues (the *HST Guide Star Catalogue*, for example).

The first CCD cameras used by amateur astronomers appeared toward the late 1980s. Their detectors had a relatively modest number of pixels (192×165 pixels for the ST4 and the LYNXX, and 377×244 pixels for the ST6). The appearance of the KAF-0400 by Kodak allowed the larger manufacturers (SpectraSource, SBIG, Meade) to make their prices more reasonable for 512×768 pixel cameras (but, unfortunately, these pixels are quite small: 9 $\mu m\times9$ μm). Larger cameras (1024×1024 with 15 to 19 μm) are still very costly for amateur astronomers, but the situation should improve over the next few years. Meanwhile, larger formats (2048×2048 and 4096×4096) are, for the moment, reserved for professional astronomers.

We can compare the spatial dynamics of a CCD camera to those of photographic film. Kodak Technical Pan 2415, which is the best film for astronomy, has a resolution of 5 to 6 μm. Therefore, we can deduce that a photograph taken with a format of 24 mm×36 mm is made up of 4000×6000 points. So, from a spatial dynamics point of view, we will have to wait for 4096×4096 cameras even to rival photography.

2.1.2 *The shape and size of pixels*

As explained in the previous chapter (section 1.1.3), the preferred choice is a camera which has square or near square pixels. The ST4 and LYNXX (16 μm×13.75 μm) cameras and the ST6 (27 μm×23 μm) are acceptable with regard to their pixels, but it would be awkward if the difference between the two pixel dimensions were greater than 20–25%.

The same applies to high-precision photometric tasks (see section 1.1.3). Here one should avoid CCDs that have a dead zone between pixels (the Thomson 7852, for example). Then again, this dead zone does not hinder images of objects (planets, nebulas, etc.).

The size of the pixels is not as important as their numbers. Let us compare two hypothetical arrays, A and B, with the same number of pixels and similar in every other way except that A's pixels are twice as large as B's (four times larger in area). To a first approximation, we can assume that both of these arrays have the same performance: if array A uses optics with the same diameter but with a focal length twice as long as array B, they represent the same angular resolution and cover the same area. When the same portion of the sky corresponds to the pixels of each array and when the diameter of the optics is the same, the quantity of light received by each pixel is the same; at equal sensitivity, the signal given by each array is therefore the same.

A practical difference between these two arrays can become apparent when using them for prime-focus imaging: in terms of the desired field and resolution, the telescope used is probably better adapted to one or the other array; it would be better to use array B directly at the focus rather than array A behind a Barlow lens; also, it would be better to use array A directly at the focus rather than array B with a focal reducer.

If priority is given to the field of view instead of resolution, it would be difficult to find, for a given diameter, a focal instrument short enough to be compatible with a small array, and this instrument risks displaying considerable aberrations far from the optical axis (a small F/D ratio). However, some instruments show considerable aberrations incompatible with small pixels even at the prime focus. In high resolution imaging a similar problem can occur: because it is easier to produce some enlargements than others with eyepieces or Barlow lenses.

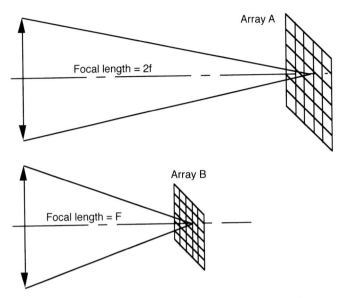

FIGURE 2.3 A schematic illustrating both A and B arrays described in the text.

Large pixels permit us to store more electrons than small ones; this allows array A to saturate less quickly than array B; it can therefore accumulate more signal on a bright object (this criterion does not apply in deep-sky astronomy where one is working far from saturation). On the other hand since the dark current per pixel is proportional to the surface area of the pixel, array A is at a disadvantage when it comes to long exposures, and furthermore it will be a little harder to cool because of its size.

As we have seen, the size of a pixel can be useful for a specific job with a certain telescope, but, in general, we should not accord it too much importance.

2.2 Electronic and thermal characteristics

2.2.1 *Read-out noise*

The origin of read-out noise During the exposure, several electrons are generated in each photodiode. At the moment of the reading, the electric charge produced by these electrons is transferred to the output amplifier of the array and then measured electronically by the camera. A number corresponding to

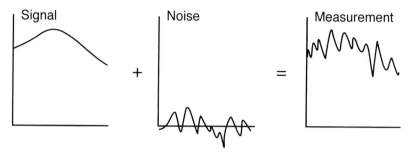

FIGURE 2.4 On the left, a one-dimensional representation of a signal. In the center, the contribution of random noise with an average of zero. On the right, the superimposition of the two components represent what is actually measured.

this measurement is sent to the computer at the memory address that corresponds to the photodiode concerned. If the array and camera electronics were perfect, the number registered by the computer would be exactly equal to the number of electrons contained in the photodiode. Unfortunately, this is never the case, and an uncertainty is attached to the count sent to the computer.

The origins of this imprecision are partly due to the CCD itself, which cannot give the electronic system the exact number of electrons read because of losses during the charge transfer and the noise at the output level. Noise is also partly due to the analog electronic system responsible for amplifying the CCD signal.

Undesirable level fluctuations around the true value constitute what we call 'noise'.

If the CCD is read too quickly, the ineffectiveness of the transfer, which is linked to CCD noise, increases. Furthermore, the electronic system is less precise when it works quickly. The camera designer must choose a reading rate that takes into account both the desired quality of the image and the impatience of the operator to see the image appear. While a photosensor is read, the charge contained in the output of the CCD is constant, but if we examine the values measured by the electronic system, at the end of the analog sequence, we notice continuous variations: analog noise. This is the electronic components themselves generating noise.

To obtain the most exact measurements possible, an electronics engineer makes use of several tools: multiple measurements, signal integration, filtering, etc. Signal filtering is universally used, but it is too harsh and may even interfere with the signal being measured. In terms of the origin of the noise, the electronics engineer must choose some frequencies that will be rejected by the filter

FIGURE 2.5 The subtraction of two images taken by one CCD camera in total darkness with a short exposure (precharged or biased image). With the absence of read noise, this image should be uniformly black. Therefore, we can ascertain that the pixels represent different and random levels of gray, characteristics of read noise. The contrast of this image was greatly increased to render this phenomenon visible.

(those that are the noisiest), and some that will be measured (those that are most characteristic of the signal).

The choice of the components is also very important. For certain strategic parts, it is necessary to choose components of very high quality that generate very little noise. But opting for the best quality everywhere would lead to extremely expensive electronics. Also it should be seen to that the most sensitive components stay at the lowest possible temperature: the electronic noise of a component is reduced by a factor of 2 each time the temperature is lowered by about 7 °C.

Optimization of the electronic system is not simple and the choices are endless. It is the ability of the electronics designer (and the prices of the components) that contribute to a camera's quality.

The influence of read-out noise The electron is often used as the unit to describe read-out noise. Therefore, its value can easily be compared to the signal furnished by the CCD or to the maximum capacity of the photosensors. Very good cameras, such as those used in professional astronomy, have a read noise less than a dozen electrons per pixel. Cameras geared toward amateur astronomers have, in general, a read noise of about 10 to 100 electrons per pixel. Anything over a few hundred electrons begins to hamper the camera's abilities for astronomical observation.

In practice, noise is caused by an error in the signal's measurement contained in each pixel. This error is completely uncorrelated from one pixel to another and one image to another.

Read-out noise is one of the principle limitations to the detection threshold of the CCD cameras, as well as the quality of their images. It is one of the fundamental criteria when choosing a CCD camera.

Some authors recommend, in certain cases, to replace a 10 minute exposure, for example, with ten 1 minute exposures (in the case of a comet's image for example, to avoid producing a comet trail with the telescope), and contend that the 'total' image quality obtained is the same because of the linearity of the CCD. Actually, if a 1 minute exposure gives us signal S, the sum of ten 1 minute exposures or one 10 minute exposure both equal an identical signal of $10S$. But we cannot ignore the read-out noise that occurs only once in the 10 minute exposure and 10 times in the ten 1 minute exposures. Luckily, because of its random patterns and its lack of correlation from one image to another, the law of averages takes effect, so well, in fact, that the noise of the sum of N images is not equal to the noise of an image multiplied by N, but only by $\sqrt{N}$. The quality of an image is usually defined by its 'signal to noise' reading (usually denoted S/N). In the previous example, the total noise of the ten images of 1 minute would be three times weaker than the noise of one 1 minute exposure, but three times stronger than the noise of one 10 minute exposure.

2.2.2 *The thermal load and noise*

As we have seen in section 1.1.5, a CCD constantly generates electric charges independently of any light. These charges are called 'dark currents' or 'thermal electrons', reminding us of the important effect that temperature has on the CCD: the number of these charges' are reduced by a factor of 2 each time the CCD's temperature is lowered by about 6 °C.

This phenomenon is reproducible, which allows us to subtract an image of the average dark current. But for each pixel of every image, the number of thermal electrons can only be known to plus or minus the square root of that number on average. Therefore, we cannot subtract out all of the thermal electrons contained in a given pixel: there will always remain a certain charge whose value, on average, is equal to the square root of the number of thermal electrons, which we do not know if they represent noise (in which case they would have to be removed) or if they are created by illumination of the CCD (in which case, they should not be removed).

Thermal current has two consequences:

- It fills the photosensors with unwanted charges which we just about know how to keep count of after reading the image, but these charges lead to saturation if the exposure is long enough.
- It cannot be perfectly corrected for, and thus creates a thermal noise whose value is equal to the square root of the number of thermal electrons.

We can successfully limit the dark current by firstly choosing an adequate CCD (MPP technology, for example), and secondly ensuring effective cooling of the detector.

CCD camera manufacturers provide in their documentation the temperature reached by the CCD during its operation as well as the dark current. The latter is generally given in electrons per second per pixel.

As in the case of the read-out noise measurement, the values provided by the camera manufacturers are sometimes a little optimistic. These values are interesting to compare with the values published by the makers of the CCDs used. The temperature recommended for the CCD must be carefully noted: effectively, a cooling system with a Peltier module base provides an almost constant difference of temperature between the detector and its radiator. The radiator (if it works by convection) is inevitably hotter than the ambient air; therefore, the CCD's temperature depends strongly on the ambient temperature. Hence, it is easy to claim a remarkably low CCD temperature when measured against an ambient temperature of 0 °C instead of 20 °C or using the radiator's temperature as a reference rather than the air's (a difference of 10 °C, therefore, would not be surprising).

The value of the dark current is, for an astronomical camera, just as important a criterion to consider as the read-out noise value.

The dark current value is in the order of one to a few electrons per second per pixel (sometimes less) for high-end cameras. The value can reach a few hundred electrons per second in cheaper cameras. Anything over a few hundred electrons per second compromises the camera's use for long exposures.

To evaluate thermal noise, we must multiply the value of the dark current by the exposure time and take the square root of the result. It is wise to compare thermal noise to read-out noise.

For example, the LYNXX 2 camera's specifications list a read-out noise of 20 electrons and a dark current of 50 electrons per pixel per second. For a short planetary exposure of 0.5 seconds, for example, the thermal noise will be $\sqrt{50\times0.5}=5$ electrons; the read-out noise, in this case, is predominant. On the other hand, for a long exposure of 450 seconds, for example, the thermal noise will be $\sqrt{50\times450}=150$ electrons, which is clearly above the read-out noise.

2.2.3 *Sensitivity to interference*

Not all noise sources are caused internally within the camera. The video signal coming from the CCD is so weak (we can measure a few electron charges) that little is needed to produce interference. In the real world, there is no shortage of radio interference sources, starting from the very computer that controls the camera. A badly shielded power supply or a poor ground is all that is needed to have a CCD image streaked with obvious parallel lines. The Pic du Midi Observatory has the harshest of radio environments: their CCD cameras are condemned to be neighbors to a powerful television transmitter of 15 kW.

The CCD camera designer must pay attention to its sensitivity to interference, and this implies some non-negligible constraints (amplifier stages in the body of the camera although one wishes to limit the number of components at that location, the shortest wire connections possible, shielding, etc).

Unlike components used for aircraft electronics (which, admittedly, are very expensive), CCD cameras are not subjected to interference susceptibility tests. The manufacturers do not inform their customers about this, and the potential buyer must rely on the trials and tribulations of other users.

2.2.4 *Levels of digitization*

After being amplified, the video signal from the CCD must be digitized to be understood by the computer (see section 1.2.2). The analog-to-digital converter (also called the ADC or digitizer) is responsible for this task. It is this electronic circuit that codes the analog signal with a specific number of binary variables called 'bits'. A coded signal with N bits can take 2^N different values. This signal is said to be coded on 2^N levels of gray (referring to a black and white image), also called ADU, by electronics engineers.

Frequently used digitizers in CCD cameras code information in 8 bits (256 levels of gray), 12 bits (4096 levels of gray), 14 bits (16384 levels of gray) or 16 bits (65536 levels of gray).

FIGURE 2.6 This image of Messier 42 was obtained during a CCD observing session at Bonascre with a LYNXX camera coupled onto a 180 mm $F/D = 3.5$ telephoto lens with a 15 second exposure. The image is viewed with a gray background to highlight the oblique lines caused by interference from the computer feed which was not sufficiently shielded. Reference: Mathieu Sénégas.

Levels of digitization are an important criterion, since astronomical images need a large dynamic range.

For example, the detail that can be seen on Jupiter represents, in relation to neighboring bands, a contrast less than 1/256th of the luminosity of the brightest point on the planet; the exposure is adjusted so that no area is saturated; the brightest point does not exceed 256 levels of gray if the camera has an 8 bit digitizer; in such conditions, low-contrast data will be attributed the same level of gray as its environment and will go unnoticed.

An 8 bit converter only represents 256 levels of gray, which is an inadequate dynamic range for a great deal of astronomical images, except when doing automatic guiding. There are some very reasonably priced 8 bit cameras on the market. These models should be avoided if one wishes to produce CCD images that surpass normal photography. A 12 bit converter offering 4096 levels of gray is considered a minimum to produce astronomical CCD images worthy of the name.

The Thomson 7895 CCD detector has a capacity of 450 000 electrons per pixel. A camera aiming for the high-end range must have a read-out noise around 10 electrons or less. The capacity of the CCD, therefore, is 45 000 times that of the read-out noise. Since the ADU must be less than the noise, so we can properly analyze an image, we can see that a 16 bit converter, with its 65 536 levels of gray, is just sufficient for this camera. A high-end camera using the Thomson 7895 would not be adequate with a 14 bit converter.

The opposite example also exists: such a commercially available camera equipped with a 16 bit converter uses a Texas Instruments CCD which has a maximum capacity of 200 000 electrons per pixel and a read noise of 120 electrons (not including the other electronics). A 12 bit converter would be sufficient for this camera; the 16 bit converter would not benefit it in any way – unless it is on special offer!

Note also that a low-end 16 bit converter is barely as good as a competent 14 bit one. The results look good for the 16 bit (and in the camera vendor's showcase too!) but the 16th bit, and sometimes the 15th, do not have significant value.

It would also be interesting to know how the reference level is removed from the signal while the pixel is being read. But this detail is not generally provided by the manufacturers. If the subtraction is made numerically after the conversion on the one hand of the signal and on the other of the reference, then the error is rounded off owing to the digitization being done twice per pixel, which would make such a camera, if it were using a 14 bit converter, act like a 13 bit one.

The speed and accuracy of a converter must be adapted to the camera's performance. In fact, the higher converter's bit number is, the slower it is, or at equal speed, the more expensive it is! Luckily, prices tend to drop, for the same performance, but a fast 16 bit converter is still an overt luxury reserved for the high-end cameras.

It is for this reason that we can find 18 bit converters in only very high-end and very expensive professional cameras. The transition from 16 bit to 18 bit also presents an additional problem. Since computers work by bytes, that is by 8 bit groups, the data coded on 12, 14, or 16 bits need, from the computer, the manipulation and storage of two bytes per pixel; the transition to 18 bits requires three bytes per pixel.

In conclusion, the number of bits of a converter is an important criterion for a CCD camera: 12 bits seems necessary for a basic camera and 16 bits for a high-end camera. But looks can be deceptive: the rest of the electronics should be checked to see if they are up to the converter's capabilities.

2.3 Photometric characteristics

2.3.1 *The capacity of pixels*

Pixel capacity is not an important criterion to consider when choosing a CCD camera, but it is interesting to examine in terms of what role it plays in the camera's detection, as we can see below. In figure 2.7 the large pixel CCDs are at an advantage since their capacity is generally greater than the small pixel CCDs (see section 1.1.5).

In general, one should avoid producing images with pixels that are too

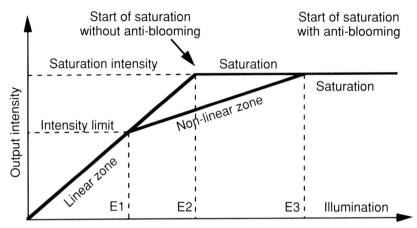

FIGURE 2.7 The response curve of a Texas Instruments TC 241 CCD, expressed by the output intensity as a function of the light received. When the anti-blooming device is not activated, the component responds linearly until an illumination E_2, then it saturates. When the anti-blooming device is activated, the component is linear only until an illumination E_1 (lower than E_2). The intensity limit's value is between 56% and 90% of the saturation intensity value. Reference: Texas Instruments documentation.

close to saturation because the linearity of the CCD is not as good near the limit and this leads to an unsightly blooming effect.

One should also be careful of CCD systems that offer a 'clocked' anti-blooming device: when activated, the CCD can lose its linearity on, roughly, the upper half of its capacity, rendering bright object images photometrically useless (see figure 2.7). More modern forms of 'vertical' anti-blooming systems are better for linearity and can be good to 90% of saturation.

2.3.2 Sensitivity and detection thresholds

The sensitivity of a CCD is characterized by the number of electric charges generated for a given illumination. The quantitative output equivalent (see section 1.5.1) is an expression of this sensitivity.

Modern CCD detectors differ very little from one another in terms of their global sensitivity: the maximum quantitative output varies between 40% and 60% according to models. More noticeable diferences can appear at certain wavelengths, at the extremes of the spectrum in particular.

We shall not mention the back-lit CCDs that have a better ultraviolet sensitivity, since their prices are not, for the moment, geared toward amateur astronomers.

Sensitivity, which is one of the primary criteria when choosing a photographic film, does not have such a fundamental importance when it comes to CCD cameras. Detection threshold is a less intuitive notion than sensitivity. It characterizes the camera's capacity to make low-contrast objects appear on an image. For example, the detection of a very faint star against the sky or very low contrast planetary detail.

Since the CCD is linear, the object to be detected, even if it is faint, is always found on the image with a certain signal: the pixel that received the weak star's image generates electric charges induced by the star's light. But the image is noisy, which means that from one pixel to another there are charge variations that do not correspond to a lighting variation. The faint star will only be detected if we know how to recognize signal variations provoked amid the noise linked fluctuations, and only if the signal is clearly stronger than the level the noise can reach. Given that, it will only be detected if the signal is three times the average fluctuation of the noise (which means a signal to noise ratio of S/N equal to 3).

Detection threshold always comes back to comparing a signal's intensity induced by a faint object (the signal intensity is directly related to sensitivity, and therefore it will vary very little from one CCD to another) to the noise of the image. The electronic noise, therefore, is a determining factor in a camera's detection threshold.

Two factors contributing to image noise are fundamental characteristics of the CCD camera: read noise and thermal noise. A third factor is completely independent of the camera: photon noise.

Photon noise is created by the observed light source itself. Let us take, for example, the detection of a faint star in a faintly lit sky. The luminosity of the sky is taken to be constant and uniform. If this were really the case, it would be easy to remove it from the CCD image to reveal the star's image super-imposed against the sky. Unfortunately, the sky's 'homogeneous' luminosity undergoes random fluctuations, expressed in number of photons. The average value of these fluctuations is equal to the square root of the average number of photons emitted by the sky. The resulting noise very much resembles thermal noise: both are the square root of unwanted generated, but globally reproducible electrons; they are both proportional to the exposure time.

With all noise sources being uncorrelated, the total noise of the image is on average defined as the square root of the sum of the squares of all the elements' noise:

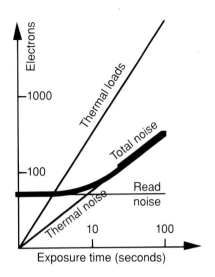

FIGURE 2.8 This diagram shows how the generation of electronic noise caused by thermal and read noise evolves as a function of exposure time. For short exposures, the noise is dominated by the CCD's read-out noise. For long exposures, it is the thermal noise that dominates. This remains true in the case of deep-sky objects where photon noise remains weak. In contrast, for planetary imaging, photon noise generally dominates, regardless of the exposure time.

$$B_T = \sqrt{B_r^2 + B_{th}^2 + B_{sk}^2}$$

where the total noise is B_T, the read-out noise is B_r, the thermal noise is B_{th} and the sky noise is B_{sk}. Therefore, we know that B_{th} and B_{sk} are equal respectively to the square roots of the number of electrons produced by the dark current and the light from the sky. If C_1 and F_1 are the number of charges produced in 1 second by the dark current and by the light of the sky and t is the time of the exposure, then

$$B_T = \sqrt{B_r^2 + (C_1 + F_1) \times t}.$$

In this equation of noises, we have included read-out noise and thermal noise only once. So, if from the image taken of the sky we were to directly remove a dark image produced at the same temperature and with the same integration time in order to subtract out the dark current, the read noise and thermal noise appear twice: once in the image of the sky and once in the dark image. Therefore, the total noise value is

$$B_T = \sqrt{2B_r^2 + 2B_{th}^2 + B_{sk}^2}.$$

However, some modern software allows the adjustment of these dark images produced through very long exposures or averaged from a large number of dark images. Under these conditions, if the work is well done, we can consider the dark image noise to be negligible, in which case the thermal noise is only counted once.

If we admit that a signal is only properly detected once it is at least three times the average value of the total noise B_T, a better detection threshold will result from minimizing this noise. This is one of the major worries of both the designer and the CCD camera user. For a short exposure, it is understood that the read-out noise will be predominant. For a long exposure, the total noise depends strongly on the value of the dark current and the sky's signal. Of these terms, one or the other could be decisive, after the performance of the camera (above all, the cooling efficiency) and the sky brightness (due to light pollution and the presence of the Moon). The F/D ratio of the telescope also plays a role: the faster the telescope is, the more light will illuminate each pixel.

A CCD camera allows us to produce images of faint objects under a full moon or from the heart of a large city under a sky lit with light pollution. Under these conditions, photography is not possible because the film would fog. As for CCDs, we can remove the background fog from an image, even if it is close to saturation, to deal with the signal coming from the detected object. We can even produce a dozen identical images, remove the background fog on each one and then combine them to obtain more signal, which is very tricky to do in photography. The limitations of CCD images, however, come from the noise of the light interference we are trying to remove.

If one is buying a CCD camera to always use under polluted light conditions (in the center of a large city, for example) and with a large-aperture telescope, it is unreasonable to choose a high-end model whose dark current will be very weak, since the sky brightness provides the same limitation as the dark current.

What is the ultimate detection limit of a CCD camera? We will do the calculations using, for example, a LYNXX PC basic camera. This camera's maximum pixel capacity is 150 000 electrons.

We will use an exposure time such that the dark current and the sky brightness fill the photodiodes to a value near saturation, say 120 000 electrons (the photon noise will be $\sqrt{120\,000}$, that is 346 electrons). The electronic noise of this camera (80 electrons) is just a small contribution to the images total noise ($80 \times 80 = 6\,400$, small compared to $346 \times 346 = 120\,000$). The total image noise with pixels filled to 120 000 electrons will be $\sqrt{126\,400}$, say 355 electrons. How much time will it take for the camera's pixels to be filled to 120 000 electrons? To simplify, let us suppose that the image is taken under a perfectly black sky; the interference charges, therefore, come almost entirely from the dark current, which is 100 electrons per second per pixel for this camera, at an ambient temperature of 20 °C. We would therefore need an exposure of 1200 seconds (20 minutes) to fill the pixels to 120 000 electrons.

Within these 1200 seconds, the weakest star detected would have to provide a number of electrons per pixel equal to about three times the total noise, or 1000 electrons per pixel. Keeping track of the quantum efficiency of the LYNXX camera (on average 40% for the spectrum of a G-type star), a G-type star of

FIGURE 2.9 A CCD image of a 14th magnitude galaxy NGC 2793, taken with the
60 cm telescope at Pic du Midi with a LYNXX camera, by Patrick Martinez. Exposure:
14 minutes under a full moon. The background noise of the image mainly comes from
the sky's illumination, but this galaxy would have been inaccessible using photography
under the same conditions. CCD images taken with this telescope and the LYNXX
camera under a full moon have allowed the detection of 19th magnitude stars in 50
seconds – a photographer's dream.

magnitude 20 provides 0.016 electrons per second per cm² of collecting surface.
Taking a 20 cm diameter telescope with a factor of 0.2 obstruction and a
reflectivity of 0.92, this star will supply the detector with 4 electrons per second,
or 1 electron per second and per pixel if we suppose that the star's image is uniformly spread across 4 pixels.

In 1200 seconds, this magnitude 20 star will supply 1200 electrons per pixel,
or about three times the value of the total noise per pixel; hence, it will be
detected.

The above calculation allows us to establish that a LYNXX camera mounted
on a 20 cm diameter telescope can attain a magnitude of 20 in a 20 minute
exposure. To improve on this result, we would have to:

- either cool the CCD even more: for an ambient temperature of 12 °C, for
  example, the thermal current would be about 2.5 times weaker, and
  therefore, it would be possible to increase the length of exposure by 2.5
  times (supposing that the sky brightness continues to remain negligible!); hence, stars one magnitude fainter could be detected with an
  exposure time of 50 minutes;

- or produce several exposures of 20 minutes that can be combined after
  removing the dark current; the gain in signal to noise ratio is the square

root of the number of exposures: we would then have to produce 6 successive images of 20 minutes to gain one magnitude of detection.

The detection threshold is linked to the total capacity of the pixels: if camera A has a capacity twice as large as camera B, all other factors being equal, it allows a maximum exposure twice as long; the signal to noise ratio is therefore improved by a factor of $\sqrt{2}$ which corresponds to a magnitude limit gain of 0.4. In reality, a superior pixel capacity generally implies larger pixels, which, with equal technology and identical cooling, produce a larger dark current per pixel; there is therefore no simple rule of thumb.

2.3.3 *Photometric precision*

The linear response of the CCD detector together with a high sensitivity make it an excellent photometric detector. The number of electrons generated within the pixels corresponding to the star's image are proportional to the light received; therefore, it is sufficient to measure this electric charge to directly obtain a luminosity measurement. The measurement's precision, therefore, is linked to the quality of the camera but more specifically to the intensity of the electronic noise.

Photon noise In addition to the camera noise, we can add photon noise coming from the star, which we have not kept track of in our calculations because it did not influence matters significantly; the photometric precision, however, is otherwise. Like all light sources (such as the sky in the detection calculations, for example), the observed star sends a number of photons whose average number defines the star's brightness, but this number undergoes statistical fluctuations whose value is once again equal to the square root of the number of photons received. This has nothing to do with the possible variable behavior of the star, but is linked, rather, to the random way the star emits light. If we consider the total number of photons emitted from a star, the number is so huge that the relative value of statistical fluctuations (that is, the square root of the number of photons divided by the number of photons) is completely negligible and we can consider the emission of the star to be strictly constant. On the other hand the number of photons from the star detected a few dozen or a few hundred light years away from it by a tiny amateur telescope is not too high and statistical fluctuations must be taken into account. The photometric precision of a would-be perfect camera would still be limited by photon noise from the star being measured.

In making the hypothesis that the bulk of the light falls on a single pixel (which is paradoxically an unfavorable situation in this calculation!), the integration time would be adjusted so that the star's light could create a number of electrons close to saturation. Let us take, for example, a camera whose pixels saturate at 120 000 electrons; if the measured star generated 90 000 electrons, the photon noise would be equal to $\sqrt{90\,000} = 300$ electrons; for comparison, this value is equal to 1/300th the signal of the star (300/90 000), which corresponds to an uncertainty of 4/1000ths of a magnitude. A pixel four times larger would give results twice as good. By contrast, a star emitting a weak signal (which is unfortunately often the case) sees its relative contribution of photon noise rise: a star that generates 900 electrons produces a noise of 30 electrons, or an uncertainty of 1/30, or 0.036, magnitude.

The evaluation of photometric precision We will evaluate, as the heading suggests, the photometric precision of the set-up examined in the previous section: a LYNXX PC camera mounted on a 20 cm telescope, aimed at a 10th magnitude star. This star provides 10 000 times more electrons than the 20th magnitude star previously examined, or 40 000 electrons per second. The integration time will be 3 seconds to collect 120 000 electrons, compatible with the camera's capacity.

To be sure of measuring all of the light from the star, we will add the received signal onto a group of several pixels centered on the image of the star; the measured surface depends on the quality of the image obtained; here we take, for example, a 5×5 pixel square, or 25 pixels. But a measurement done this way not only contains the light of the star but the light of the sky on the measured surface. To remove the sky's contribution, we would have to measure it close to the star but not so close as to include light coming from the star being studied (or another star). Also, we would finally take a 7×7 pixel square whose center square of 5×5 pixels would contain the star's measurement and the square annulus 2 pixels wide the sky background.

As we will measure in the standard 49 pixels (7×7), the final result will be perturbed by 49 samples of read-out noise, 49 of thermal noise, and 49 of sky noise, to which we will add photon noise from the star. All these noises being uncorrelated, the total noise is equal to the square root of the sum of the elementary noises:

$$B_T = \sqrt{49B_r^2 + 49B_{th}^2 + 49B_{sk}^2 + B_{ph}^2}.$$

B_T is the total noise, B_r is the read-out noise of one pixel, B_{th} is the thermal noise of one pixel, B_{sk} the background sky noise of one pixel, and B_{ph} the photon noise. C_1 and F_1 will be the number of charges produced in 1 second by the dark current and by the sky's light on one pixel; then P_1 is the total charge created by the star in 1 second, t is the exposure time and as in section 2.3.2,

$$B_T = \sqrt{49B_r^2 + (49C_1 + 49F_1 + P_1)t}.$$

If we were to take a read-out noise of 80 electrons, a dark current of 100 electrons per pixel per second, a sky brightness of 20 electrons per second per pixel, and a signal of 40 000 electrons per second, we would find a total noise of 672 electrons. The signal being 120 000 electrons, the precision obtained is about 0.5% or 0.5 hundredths of a magnitude. This precision is excellent; it is understood, however, that it will not be as good for a much fainter star, which would need a longer integration time, and hence a larger contribution from the dark current and the sky brightness.

Some modern image manipulation software has photometric measuring functions that can improve the image's precision. These functions compare the measured flux with the theoretic profile of a star's image. The adjustment produced permits the correlation of one pixel to another, hence limiting the influence of noise.

2.3.4 *The precision and linearity of the converter*

In all photometric calculations, one has to keep track of rounding errors coming from the camera's digital to analog converter. Strictly speaking, the converter introduces a supplemental noise in addition to the preceding noises, evaluated as $p/\sqrt{12}$, or about 0.29p, where p is the number of electrons per ADU.

This quantization noise is important for an 8 bit coded camera: if the full scale of quantization is 150 000 electrons, each ADU therefore covers 586 electrons, from which there is a quantization noise of 169 electrons. Then again, if the same camera is equipped with a 12 bit converter, each ADU corresponds to 17 electrons, and hence, a noise of only 10 electrons.

Cameras used in astronomy, for photometric measuring purposes, must be equipped with at least a 12 bit converter. In general, the ADU must represent a smaller number of electrons than the read-out noise of the camera.

Since the digital to analog noise is 0.29 times the value of an ADU and each noise contributes its square for the total noise calculation, the digital to analog noise's contribution should not exceed one tenth of the read-out noise, for a well made camera. There is another possible difficulty with the converter: it sometimes produces a non-perfect linear response. This means that even if the full precision of the converter can be guaranteed for the comparison of two nearby bright objects, the precision is degraded if we compare measurement taken from one end to the other of the intensity range (between a bright star and the sky background, for example); the loss of precision is generally in the order of 1 bit. For example, certain converters that claim to be 16 bit (and actually render 16 bit performance on planetary images) do not have more than 15 bit precision over all their intensity range, because of linearity errors (and will only give 15 bit performance for photometers).

2.4 CCD quality

It is extremely difficult to produce a CCD detector without any faults, particularly for large arrays. Also, manufacturers test the CCDs one by one as they leave the factory so that they can evaluate individual faults for each unit. This also dictates different 'classes' or 'grades' which define the CCD's quality . . . and their price!

The principle faults found in an array are:
- Hot pixels that saturate very quickly.
- Dead pixels that do not create a charge.
- Gray pixels, whose sensitivity differs noticeably from the average pixels of the array.

These faults can affect isolated pixels, groups of adjacent pixels, or entire columns. Once we are dealing with a column or line of gray pixels, we call it a 'cold' line or column.

Kodak gives precise definitions of the types of faults of their arrays:

- *Point defects:* pixels whose response differs by more than 6% from adjacent pixels whilst they are illuminated to 70% of saturation. 'Adjacent pixels' means a square of 100×100 pixels centered on the pixel in question.
- *Cluster defects:* a group of a maximum of 5 adjacent defective pixels.
- *Column defects:* a defective group of pixels located in the same column.

For each CCD class, the manufacturers generally publish the maximum number of pixels or groups of pixels that are defective, sometimes specifying a different number for the central zone of the array (judged more important for the user) and for the rest of the array.

In general, the prices of CCD detectors vary considerably from one quality class to another, especially for the larger arrays. This shows how difficult it is for manufacturers to approach perfection.

Luckily, for amateur astronomers, the images can undergo sophisticated processing, and modern software has functions that allow us to correct an array's cosmetic faults.

Table 2.1

KAF-0400	Grade 0	Grade 1	Grade2
Point defects	0	5	10
Cluster defects	0	0	4
Column defects	0	0	2

Class	Point Defects		Cluster Defects		Column	
	Total	Zone A	Total	Zone A	Total	Zone A
C0	0	0	0	0	0	0
C1	≤15	≤6	0	0	0	0
C2	≤30	≤15	≤12	≤6	≤6	0
C3	≤60	≤30	≤24	≤12	≤12	≤6

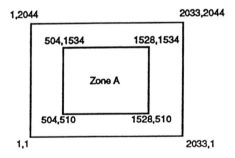

FIGURE 2.10 An extract from the documentation for the KAF-4200 CCD (2033×2044 pixels) made by Kodak. Notice that the manufacturer distinguishes defects in the central zone and in the peripheral zone.

Once a single point or column is defective, we can replace their values with an interpolation produced from adjacent pixels. Some software functions can do this very easily; the operator can even impose a replacement value in the pixel of their choice. The task is trickier when several adjacent pixels are defective. Advanced acquisition software can automatically correct pixel defects or column defects associated with the arrays they are controlling.

Thanks to such software, amateur astronomers can be content with a camera having an imperfect detector (within reasonable limits), which permits them to own respectable cameras without paying astronomical prices.

2.5 Acquisition functions

2.5.1 Exposure

Each camera is furnished with software which is indispensable to its operation, since it is controlled by a computer. This software contains the acquisition functions linked to the camera, and certain image processing and display functions.

Image processing isn't done, generally, during the observation stage; it is therefore unnecessary for the camera to have advanced image processing functions. It must simply allow the display and succinct evaluation of images so that the observer can simply control the quality of acquisitions in real time.

Image processing functions will be specifically studied in the section of this book dealing with software, which is generally independent of cameras. Here we will examine acquisition functions, that is, the functions necessary to produce images during the observation period. The presence of one function or another can be considered a plus for the camera concerned: it is also important to judge the user friendliness of the acquisition software: during the middle of the night, in the cold, perhaps in an uncomfortable position, when accumulating the maximum number of images is a priority, one is not in the mood to quibble with one's computer.

This is the fundamental function: under commands from the operator, the computer controls a sequence that ends in the acquisition of an image.

This sequence can be relatively complex: firstly, empty the CCD array of its charges just prior to the exposure, then open the shutter, time the exposure, close the shutter, the possible transfer of pixels to the memory zone, one by one read and convert pixels, place pixels in computer memory, and finally display the image on the screen after possible preprocessing.

For the observer, all of these steps must be clear. It is important that these functions be launched by a single command, so as not to overload planetary image sequences, for example, or because the observer could find himself or herself in difficult working conditions (for example, darkness, or having an eye on the eyepiece) at the moment of acquisition. After the acquisition, the image must be displayed on the screen to allow immediate control.

The operator must be able to enter the parameters of the exposure before initiating it, and appreciate that they will all be kept as defaults for subsequent exposures so that they will not have to be entered for each exposure. The essential parameter is the length of exposure, but one might also have to specify the binning rate, the size of the window selected, visualization thresholds, or automatic preprocessing before the image is displayed, and so on.

It is important to have a command that allows an interruption at any time during the exposure and not to be a slave to the computer and be obliged to wait until the end of a 15 minute programmed exposure as clouds arrive after 2 minutes, or notice that you have forgotten something important that renders the image useless immediately after opening the shutter.

2.5.2 *Automatic mode*

This mode allows the observer to take several successive exposures without having to restart the camera between each one.

These images are not saved onto the hard disk, which would quickly be filled up, but displayed on the screen. The screen refreshes each shot, which gives a sort of low-rate cinematography video at low frequencies. Using this mode is very valuable when making fine adjustments and targeting the telescope.

In its focusing function, this mode would be very advantageous when coupled with windowing, especially for arrays that take a long time to read in their entirety, so as to obtain a screen refresh as frequently as possible.

In its targeting function, it is necessary to possess a programmable option or automatic adjustment of the thresholds to locate the faintest objects marked with the shortest integration time possible. Coupling with binning mode could be equally useful during targeting.

2.5.3 *Read time*

In certain cases, it is important to quickly read the image produced by the camera; either to make some operation such as focusing and centering, less tedious or because the observation itself needs a high image rate. For example, in planetary imaging, we want to produce as many images as possible from which we select the best; the more we have to choose from, the stricter the selection, and the more we can hope to limit atmospheric turbulence degradation. Since the exposure time is very short (less than a second), it is the read time that limits the acquisition rate. Another example where a quick read is necessary is when the CCD camera is used as a fast photometer. For example, in the observation of an occultation, the temporal resolution of the phenomenon depends directly on the image rate.

The read time of the array is a criterion that can be taken into account by the CCD camera buyer to a greater or lesser extent, depending on what type of observation is going to be done.

But the fastest camera is not necessarily the best: if fast reading is achieved at the expense of the effectiveness of charges transferred to the CCD, then there is a risk that the read-out noise will be high. Another speed limitation can come from the converter: in this case it is simply a financial problem: a fast converter works as well as a slow one but is much more expensive. Finally, certain camera designs are conducive to slower read times; this is why some serial links (such as the RS232 bus), for example, for data transfers in the computer are used.

2.5.4 *Windowing*

The windowing function consists in digitizing only a portion of the image for each exposure. The goal is to obtain an increased image rate when only a portion of the field is interesting.

This is typically the case for focussing with a star; the windowing must operate in automatic mode with quick images that are displayed on the screen rather than memorized by the computer. Windowing can also be used in special acquisition modes, for example if the camera is observing an asteroid occultation: it then works with a very small photometric field, but with an elevated acquisition rate (with several measurements per second); then, whether it is the image or a photometric measurement that is done in real time, it must be safeguarded, along with an accurate timing of the image.

Windowing could be unnecessary on small arrays that are read quickly anyway; on the other hand, it is almost obligatory to have a large array.

2.5.5 *Binning*

Binning mode consists in adding the CCD charges of adjacent photodiodes to produce one pixel.

In imagery, binning is, in general, symmetrical on both axes (2×2, which means that we add the lines two by two and the points two by two in each line, 3×3, 4×4, etc), until the final result gives us square pixels (if the original pixels were square); it can be very asymmetrical in spectrography (1×10, for example) when we want to concentrate the light coming from each spectral ray onto one point.

The major inconvenience of binning is immediately apparent: the images' resolution is divided by 2 within a 2×2 binning frame, by 3 for 3×3, etc. Then again, the advantages are in part being able to read the image faster (there are four times fewer points to digitize if the binning is 2×2, for example), and in part being able to obtain a better signal to noise ratio, especially in short exposures: for example, for a 2×2 binning the contents of four photodiodes are read only once, which gives us a signal four times stronger for a single reading noise.

Binning is appealing when we want images quickly without too many worries about resolution and for the image to have a limited signal to noise ratio. This is typically the case during the target acquisition stages: the higher the image rate, the faster the field will be found and identified; we do not want to waste time in digitization or integration time, from which the weak signal to noise ratio derives. Generally, loss of resolution does not impede the target acquisition task.

Another use of binning is to detect particularly faint objects: The goal here is not to save time in the digitizing process, since the image had a very long exposure, but to make every sacrifice to gain a little signal to noise ratio.

Attention must be given to the way the camera executes the binning: it is only efficient if the addition of the pixel charges concerned is done analogically before the digitizing. But this requires a special operating mode from the sequencer that must be foreseen in the design of the camera, which is not always easy to do with some CCD detectors. The temptation is great, therefore, to do the binning after the digitizing: nothing is modified in the operation of the camera, and it is the computer which adds the data from the adjacent pixels, which is extremely easy to do. Unfortunately, there is no time advantage, since each starting pixel is read independently and the gain in signal to noise ratio is of a factor equal to the square root of the number of the added pixels instead of being equal to the number of pixels.

There are cameras that offer mixed binning: an analog binning between the lines, which is generally the easiest to analogically produce, and a digitized binning between adjacent pixels of a same line. A binning of 2×2, therefore, would give a gain time of 2 and a gain of $2 \times \sqrt{2}$ in signal to noise ratio.

The camera manufacturers do not generally give their binning formulas; only a careful measurement of its performance will teach the user the mode used.

2.5.6 *Types of shutter*

We have already seen different types of arrays and ways to close the shutter at the end of the exposure. We will re-examine these ideas since they constitute an important criterion when choosing a camera:

- An interline CCD array does not need a mechanical shutter since the passing of charges for each pixel from the sensitive zone to the safeguarded zone is very quick, faster even than the closing of a mechanical shutter; elsewhere, the electric charges do not pass through the adjacent pixels' sensitive zones, as in the case of frame transfers. This results in an advantage for interline transfer arrays because the shutter is an expensive part and sometimes the source of functioning problems.

- A frame transfer array also allows the storage of electric charges in a memory zone before reading, but to have access to this zone, the charges pass through the sensitive area of other photodiodes in the column. One could foresee using such an array without a mechanical shutter, but during the charge transfer, illuminated photodiodes continue generating electrons which are added to the charges in transit. If we produce the image of a faint object, few charges are generated during the transit time and the perturbation is negligible; the same applies for the image of a bright star, since each pixel in transit does not stay under the quasi-star-like lighting of the star for very long. Then again, some problems may arise in planetary imaging. Let us take, for example, a planet or lunar field that occupies an entire Thomson 7863 frame transfer array's field. Let us suppose that an exposure of 0.1 seconds yields 400 000 electrons per pixel. At the end of this exposure, the frame transfer of this array between its image zone and memory zone takes about 0.3 ms. During this time, the pixels that are furthest from the memory zone are subjected to light from the image zones corresponding to the photodiodes traversed. If the flux provides 400 000 electrons in 100 ms, then the interference charges accumulated in 0.3 ms are 1200 electrons. This charge only corresponds to 0.3% of the signal, but it is greater than the photon noise (630 electrons in this example) and the camera's read-out noise.

- A full frame array does not possess a memory zone. This implies that the total charges of one line must be read in between each line transfer cycle. If no mechanical shutter is used, the last pixels read remain exposed to the field's light during the entire array reading. But the time necessary for the transfer and digitization of all the pixels is far longer than the line transfer time, only taken into consideration in the preceding example. The Thomson 7883 array is a full frame array containing the same

FIGURE 2.11 This LYNXX CCD camera is equipped with a full frame array. The designer has equipped it with a mechanical iris shutter, here shown closed on the photo.

number of points as the TH 7863 (with interposed image and memory zones). But the reading of this array takes several seconds; it would be unthinkable to produce an image of the Moon with such a array without a mechanical shutter. As explained in section 1.1.4, the half-frame transfer mode allows, in certain cases (planetary images), the use of a full frame array without a shutter; but this mode has constraints (only half of the field is usable) and limitations (no bright star can be in the half-field used in the memory zone; this mode is useless for the Moon and the Sun in particular).

In conclusion:

- Interline transfer arrays do not need a mechanical shutter.
- Frame transfer arrays in most cases do not need mechanical shutters, but a shutter is recommended for planetary imaging.
- For full frame arrays, the shutter is often indispensable.

2.5.7 *Planetary image acquisition*

Planetary images are characterized by a short exposure time (less than 1 second) and a struggle against atmospheric turbulence to find the best resolution possible. Turbulence is a random phenomenon which degrades different images of a series in an unequal manner. One of the ways to improve resolution consists in taking several images and selecting the best one. The more images taken, the greater the chance of achieving a good resolution. Selecting one image out of 10 is the minimum, one in 100 considerably improves the chances, and one in 1000 could be a goal for a CCD camera user.

Photographers practice image selection on the one hand by waiting for the most opportune moment through the reflex eyepiece of the camera, which is one way to select, on condition that one reacts quickly, since opportune moments are fleeting, and on the other, by taking several rolls of film, from which only one shot will be exploited. But this selection is limited owing to the price of the films and the time needed to develop each film.

The advantages of the CCD camera are firstly the displaying of the image on the computer screen immediately after the exposure, which allows one to make real time selections, and secondly the ability to eliminate a large number of images without any cost.

The goal being to choose an image from the largest number possible, it is necessary to reduce the acquisition time between two images to a minimum. We can appreciate, therefore, a camera which has array read times and screen display times as short as possible, but also has some ergonomic features, for example:

- Starting the acquisition of the next image with a single key press.
- Saving a good image with a single command; this point is less fundamental then the latter if we save one image for 100 acquisitions. Some software allows the user to create a generic file name, and each saved image will automatically take a number in order, which is also auto-incremented; this function gives the operator some reassurance, and also having the time and date of the image automatically saved.
- Allowing the image to be displayed on the screen, which facilitates the decision to 'keep' or 'trash', by an adequate adjustment of the thresholds and possibly quick preprocessing. Some software displays a comparison of the last saved image next to the current image, which is very valuable.

These features are not gimmicks; if the camera's reading and displaying takes 4 seconds, and if the operator gives himself or herself 3 seconds to decide whether each image is to be kept or trashed, he or she would require 2 hours of observation to choose from 1000 images. Three seconds is very fast to make a decision and it would be desirable that the computer help the operator as much as possible. What is more, an error in judgment or a false move can lose the historical image the user had the privilege to contemplate for 3 seconds before hitting the wrong key on the keyboard.

Although not studied in this book, there exists, as of the beginning of 1994, acquisition software, for the ALPHA 500 camera in particular that automatically selects images, within prescribed quality criteria, itself. This software allows total reassurance: as long as the telescope tracks in a reliable way, it is sufficient to launch the acquisition software and come back a few hours later to contemplate the best images of the night!

2.5.8 *TDI mode*

This mode of functioning was already mentioned in section 1.2.1. It consists in removing the charges from photosensor to photosensor in the CCD array as the image being measured moves across the array.

The principle interest of the TDI mode in astronomy is the ability to produce images with a fixed telescope: the apparent motion of the sky makes the image slide on the CCD, but in jumping from photodiode to photodiode at the same speed, the charges created by a point in the image stay in front of the point during the whole integration time; hence, the image is tracked electronically.

TDI mode has the advantage of enabling images to be produced with a telescope whose mounting does not allow it to follow the sky's motion, as in the case of Dobsonian telescopes. A small inconvenience is that the integration time must have a well defined value; it is equal to the time it takes for the sky to cross the length of the CCD and depends, therefore, on the focal distance and the size of the CCD. With a Thomson 7895 CCD, for example, counting 512 lines of pixels of 19 μm, and a 1 meter focal-length telescope, the integration time is 134 seconds. It would take twice as long with a telescope with half the focal length.

On the electronics side, TDI mode is not difficult to produce. The principle difficulty is to very precisely adjust the time interval between the CCD line shift with the focal length: in the example of the preceding CCD, if the precision is not better than 0.2%, the image will be elongated by 1 pixel in the direction of the sky's displacement. With this precision, it is necessary that the operator proceeds at a calibration speed equal to that of the computer; in any case, does anyone know the focal distance of their telescope within 0.2%?

The trickiest stage is the mechanical mounting of the camera body on to the telescope: the direction of the charge transfer in the direction of the sky's apparent motion must be very precise. In the preceding example, an angular error of 1/512 radian leads to an image motion of 1 pixel in the direction perpendicular to the sky's displacement. But 1/512 radian is only 0.1°. The camera, therefore, must be mounted onto a micrometric device turning around the optical axis of the telescope. Such devices, today, are not made by telescope manufacturers. It is probable that many CCD manufacturers, generally more taken with electronics than precision mechanics, neglect this aspect while proposing cameras functioning in TDI mode. It is the user, therefore, who must improvise as a handyman.

TDI mode presents an intrinsically independent limitation from the quality of the camera and its mounting: it is necessary that the speed of the sky's displacement be the same for the entire field (to be correlated with the charge transfer speed): but this speed varies with declination. If $\Delta\delta$ is the difference in declination between the top and bottom of the image and N the number of pixels in the array in the right ascension direction, we see that the difference in the speed of the sky will not lead to a displacement greater than 1 pixel, only for small declinations δ, such as

$$|\tan\delta| < \frac{1}{N \cdot \Delta\delta}.$$

For example, with a TH7895M array of 512×512 pixels, covering $67'$, mounted on a 500 mm telescope, one must not exceed a declination of $6°$! The same array, placed onto a 2000 mm telescope, does not cover more than $17'$ and becomes useful in TDI mode up to declinations of $22°$.

In conclusion, TDI mode brings little to a camera used with an equatorial telescope capable of following the apparent movement of the sky. On the other hand, TDI mode opens up the field of imaging to fixed telescopes. Although useless to some observers, it will be motivation for some to change to CCD. All the same, one must bear in mind that the mechanical mounting of the camera onto the telescope is not a trivial problem to solve.

2.6 User friendliness

User friendliness and ease of use of a camera are not criteria easily evaluated. Moreover, each user can have a different perception.

The potential buyer of a CCD camera should pay attention to the following criteria:

- The weight and ease of handling of the camera body, and to a lesser extent, the electronics module; these elements vary in relation to the size and stability of the telescope supporting the camera.
- The ease of connection and number of wires or pipes that clutter up the telescope.
- The starting procedure controlling the powering-up of the Peltier modules, starting a possible pump, etc.
- The user friendliness of the software: it is preferable to have drop-down windows, dialogue boxes, and keyboard shortcuts for the principle acquisition commands, rather than entering instructions line by line.

- Ease of learning: it is more agreeable to be able to use the camera intuitively after a few minutes rather than be obliged to read a thick instruction manual before producing your first image. Once again, drop-down windows are helpful to get started.

In general, the ease of use and the performance have a tendency to oppose each other. For example, a camera under vacuum with liquid cooled Peltier modules performs better than a camera in a nitrogen atmosphere cooled by natural convection, but it requires the installation of cumbersome pumps and pipes.

2.7 Automatic guiding

Even the best telescope mounts cannot perfectly track the apparent sky motion. During a long exposure, whether by CCD or photographic, the operator must normally guide the telescope. This consists in watching, with an illuminated eyepiece, the position of a reference star in relation to the cross-hairs, and making the necessary corrections to the telescope's motors to keep the star well centered. This type of work is particularly tedious and everyone dreams of an automated way of doing it.

CCD cameras are capable of automatic guidance provided that this function was foreseen by the manufacturer.

The operation is as follows: the camera produces images from a field containing a guide star, at high rate (one image every one to a few seconds). The program immediately analyzes the image once it is read to determine the photometric center position of the star. If the position of the star has moved on one image in relation to the preceding one, a correction is directly sent to the telescope's motors.

Amid the possible solutions to produce automatic guidance, the CCD camera is certainly the most efficient and easiest to use.

But this ideal solution, however, has several requirements. Firstly, the telescope must be motorized on its two axes. Secondly, the camera's output signals (for example, closing of the relays, or TTL signals) are not necessarily compatible with the telescope's command signals; electric cables, therefore, would have to be constructed to control the motors from signals coming from the camera. Finally, the system must be calibrated so that the camera knows what amount of correction it must provide for the observed offset. One has to be aware that to control the guider, the camera must read and then empty its CCD array very often. It is therefore not possible, in general, to guide and produce the image with the same camera. The solution consists in replacing the eyepiece with the least expensive camera possible (the ST4, for example), for the sole purpose of automatic guidance, and then using a second, higher performance camera for capturing the principle image. Two modern camera models allow simultaneous automatic guidance while producing an image. The ST7 and ST8 from SBIG are equipped with two arrays in the same camera head; one of them is read regularly to obtain the information necessary for guidance, while the other array is in integration. For its part, the Alpha 500 camera has a single array, but a special command of its clocks allows it to devote half of its array to guidance (permanent reading) while the other half integrates the image.

Another problem is related to the telescope drives: the CCD camera orders a correction whose duration is proportional to the offset to correct. This implies that the guiding motor's speed is more or less constant, which is the case with the right ascension motor controlled by a variable frequency and in declination by stepping motors. On the other hand, direct current motors, often used for guiding in declination, tend to progressively accelerate. Automatic guiding, therefore, can produce problematic results. Another problem can stem from slack gears and bearings in the transmission: this can disrupt the automatic system's calibration. These defects are taken into account by the ST4 camera's automatic guidance system, by providing delay times for the two motors.

3 Image production

3.1 The camera mounting

3.1.1 *Choice of optical combinations*

In order to develop a strategy that will make you a true specialist in CCD observation, it is advisable as a first step to set up the telescope's optical assembly to obtain the desired field and resolution.

The resolution The maximum resolution we can achieve is determined by the telescope's diameter, the intrinsic quality of the images, or 'seeing', turbulence, and sampling. The first limitation comes from the phenomenon of diffraction caused by the instrument's diameter: the larger the instrument's diameter is, the better the resolution. For instance, a 12 cm diameter telescope cannot resolve better than 1 arcsecond, whereas a 50 cm telescope can reach 0.25 arcsecond. It is physically impossible to reach a better resolution at the diffraction limit of a given instrument.

The second limitation comes from the observation site's atmospheric turbulence. Unfortunately, atmospheric turbulence is often larger than the diffraction limit. We can assume that anything over 1 second of exposure time and with a diameter greater than 10 cm, the resolution limit caused by turbulence completely masks that caused by the diffraction limit. In terms of long exposures (above 1 second), classical amateur sites have seeing in the order of 5 arcseconds, with the better ones going as low as 2 or 3 arcseconds.

The third limitation comes from sampling by the CCD detector. The physical dimensions of the CCD's pixels limit the resolution by dividing the image into tiny tiles. We must therefore magnify the image sufficiently at the focal plane of the CCD in order to place a minimum number of arcseconds on each pixel. This spatial sampling is managed by the instrument's focal length and by the CCD detector's pixel size.

Spatial sampling is expressed in arcseconds per pixel (arcsec/pixel). Table 3.1 gives the sampling value, in arcseconds, in relation to the instrument's focal length and the pixel size.

In general, the formula that enables us to find the sampling value is

$$\text{sampling} = \arctan\left(\frac{\text{pixel}(\mu m)10^{-6}}{\text{focal(m)}}\right).$$

Table 3.1 *Sampling values*

Pixel (in μm)	Focal length (in meters)						
	0.05	0.20	1	2	5	10	20
5	20″	5″	1″.03	0″.52	0″.21	0″.10	0″.05
10	41″	10″	2″.06	1″.03	0″.41	0″.21	0″.10
15	1′.0	15″	3″.09	1″.55	0″.62	0″.31	0″.15
20	1′.4	21″	4″.13	2″.06	0″.83	0″.41	0″.21
25	1′.7	26″	5″.16	2″.58	1″.03	0″.52	0″.26
30	2′.1	31″	6″.19	3″.09	1″.24	0″.62	0″.31

Since the pixel's size is expressed in micrometers and the focal length in meters, the pixel size must be converted into meters, which is the reason for the 10^{-6} factor in the formula. If the pixel covers too big an angle in the sky and the stars appear square on the images, this would correspond to under-sampling. This gives us unattractive and not very photometric images. Conversely, if the pixel covers a very small angle and the stars fall on many pixels, this is over-sampling. In general, over-sampling is not detrimental unless we are trying to detect faint objects.

In practice, we want critical sampling to take place where a star's full width at half maximum occupies two pixels.

Consider a 2 meter focal length Schmidt–Cassegrain telescope. We now place, at its focus, a LYNXX camera equipped with rectangular pixels: 16 μm by 13.75 μm. According to the above table, the sampling is about 1.5 arcseconds.

For deep-sky studies, we generally use a sampling of 1–4 arcseconds per pixel. For high-resolution observation, we use a sampling of 0.1–1 arcsecond per pixel. We note that with classic amateur instruments, focal lengths greater than 5 meters are obtained by adding a Barlow lens or by replacing the eyepiece with a more powerful one.

The field of view The field covered by the CCD depends on the CCD array's dimensions and the instrument's focal length. If we know the pixels' size and how many there are along a side of the array, then it is enough to multiply these two values to know the array's length of side. The field covered by the CCD will be more extended as the array's sizes increases and the instrument's focal length decreases.

Since the array has two dimensions, the field width on each side of the CCD must be determined. In the general case, the formula that allows us to calculate this field is

Table 3.2 *Field values*

CCD	Focal length (in meters)						
side (in mm)	0.05	0.20	1	2	5	10	20
2.5	2°.9	43'	8'.6	4'.3	1'.7	51"	26"
5	5°.7	1°.4	17'.2	8'.6	3'.4	1'.7	51"
7.5	8°.6	2°.2	25'.7	12'.9	5'.1	2'.6	1'.3
10	11°.4	2°.9	34'.3	17'.2	6'.9	3'.4	1'.7
12.5	14°.3	3°.6	42'.9	21'.4	8'.6	4'.3	2'.1
15	17°.2	4°.3	51'.5	25'.7	10'.3	6'.9	2'.6

$$\text{field} = \arctan\left(\frac{\text{side of CCD(mm)}10^{-3}}{\text{focal length(m)}}\right)$$

Table 3.2 above gives the field values, in arcminutes, in relation to the instrument's focal length and the dimensions of the array.

It is always difficult to reconcile a large field and a good resolution. The art of choosing an optical system consists in obtaining the largest field possible while keeping a reasonable resolution.

Consider, for example, a 2 meter focal length Schmidt–Cassegrain telescope. At its focus, we place a LYNXX camera with rectangular pixels: 16 μm by 13.75 μm. The number of pixels on each axis is $N_x = 166$ pixels and $N_y = 195$ pixels. Firstly, we calculate the length of the sides of the camera: $L_x = 16$ μm $\times 165 = 2.68$ mm and $L_y = 13.75$ μm $\times 195 = 2.68$ mm. The array, therefore, is a 2.7 mm sided square. With the help of table 3.2, we deduce that the field covered by the detector is about 4.6×4.6 arcminutes. The sampling is about 1.5 arcsecond. This configuration is suited for deep-sky observation.

In this example, the observed field easily permits the observation of most deep-sky objects, except large objects (M31, M42, M27, M51, etc.). On the other hand, it would not be easy to find the field of objects to observe, even with the *Uranometria 2000* atlas. With the help of a focal reducer, we can extend the field but risk under-sampling the image: there is a compromise though. We will see in section 3.2 that the solution consists in equipping the telescope with very precise setting circles or using a larger array camera.

If we add a Barlow lens to the previous configuration, we would have a 2.3'×2.3' field. In this case, this configuration is suited for planetary surface observation, with a sampling of about 0.8 arcsecond. We also see that the field allows the observation of all planets without difficulty.

FIGURE 3.1 This ST4 CCD camera is directly mounted onto the aperture of a rotatable eyepiece of a 180 mm diameter telescope. Unlike simpler models, this one, being equipped with a double screw, avoids the major drawback of field rotation when one proceeds with the focusing. Documentation: Vincent Letillois.

3.1.2 *Mechanical integration*

The camera body, equipped with its CCD detector, has the same role as a photographic camera with film. We must therefore be able to attach it to the telescope as is done with a regular camera.

Several mechanical interfaces exist and are more or less practical in terms of the optical assembly we wish to produce. Some cameras also accept several different interfaces; they are shaped in the form of rings that must be screwed onto the front face of the camera. The simplest adaptor system consists of fixing onto the camera body a tube whose exterior diameter is the standard eyepiece diameter (31.75 mm or 50.8 mm). The camera, therefore, is mounted as an eyepiece in the eyepiece-holding tube. This mounting is easier to produce and does not need very many accessories. It is well adapted to images taken at the telescope's focus, without intermediate optics. This technique, however, encounters several problems: if the focusing is done by rotating the eyepiece, we can no longer correct images by flat fields because of the field's rotation. Also, it becomes impossible to correctly install a filter, a Barlow lens, or a focal reducer in front of the camera.

If we wish to include an optical system (eyepiece for image enlargement, Barlow lens, focal reducer, etc.) between the telescope and the camera, it is usually necessary to use a specific mechanical component that accepts the optical system while acting as a mechanical link between the telescope and the CCD camera. But these mechanical components already exist: they are marketed by telescope manufacturers so that a photographic camera can be fixed onto these optical mountings. Each manufacture of photographic equipment chooses to make their cameras with a bayonet mount incompatible with other brand name. The standard method of attaching photographic cameras is the T connection: this is a 42 mm threading with a pitch of 0.75 mm. There exists, for the main photographic equipment brand name, connections that fit onto the camera's bayonet that adapt it to a standard female T thread. The telescope manufacturers, therefore, have marketed mechanical connectors that can be mounted onto the telescope's eyepiece drawtube, which allows an optical interface to be attached and presents, on the photographic camera side, a male T thread allowing the mounting of any camera equipped with a T conversion ring. Some CCD camera manufacturers foresaw this and equipped their camera heads with a female T threading, which allows them to be mounted like a photographic camera at the telescope's standard drawtube.

A trickier problem is mounting the CCD camera not only behind a telescope, but behind a photographic lens. This mounting can be of interest if we wish to produce wide-field images of the sky, by fixing the CCD camera, equipped with a photographic lens, parallel with the telescope. Although we can easily find the conversion rings allowing the transformation of a camera's bayonet into a 42 mm female thread, it is extremely difficult to obtain a symmetrical connection, which transforms a lens' bayonet to a 42 mm diameter male thread. The other problem is that the distance between the photo lens' focus and its mechanical fixing varies from one brand to another; consequently, the CCD camera which accepts different brand photographic lenses must use a set of rings of different thicknesses according to the brand name of the lens used.

Using lenses of different brand names is not an easy task for the camera designer: the space available between the lens' position and its focal point is quite limited.

Therefore, we must place, in this space, the connection rings, the airtight window, a space large enough between this glass and the CCD to afford sufficient thermal isolation, possibly a shutter, and, if possible, a filter drawer.

A word of warning: the distance indicators on the focusing ring of a lens mounted onto a CCD camera do not necessarily correspond to reality! Just because the lens' ring is set to infinity does not mean that images of the sky will be focused on the CCD.

3.1.3 *The chromatic aberration of refractors*

Telescopes with lenses (such as those used by astronomers) are called refractors since the light path through the glass is modified by refraction. In contrast, telescopes using mirrors are called reflectors since the light is reflected by the mirror's surface.

Unlike reflectors, refractors produce chromatic aberration, which means that the images are not exactly focused on the same plane for different wavelengths. As a consequence, accurate focusing cannot be done simultaneously for every color (see figure 3.3).

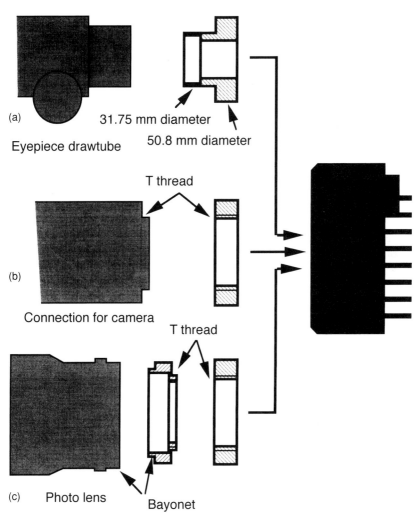

(a) Eyepiece drawtube 31.75 mm diameter 50.8 mm diameter

T thread

(b) Connection for camera

T thread

(c) Photo lens Bayonet

FIGURE 3.2 The different possibilities of mechanical mountings for a LYNXX camera. (a) A connector with 31.75 mm and 50.8 mm exterior diameters allows the camera to be directly fixed onto the eyepiece tube to be used at the telescope's focus without optical intermediary. (b) The T ring allows the camera body to be screwed onto all the connections intended for photographc cameras. (c) A conversion ring screwed onto the T ring allows the use of photographic optics, but this requires the use of a different thickness T ring for each brand of camera in order to guarantee the focusing.

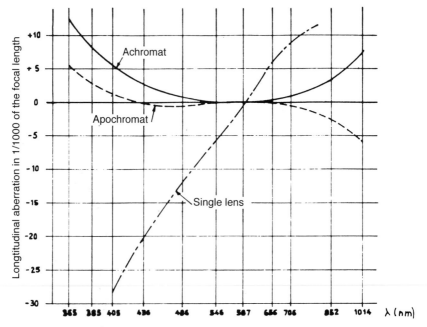

FIGURE 3.3 This graph represents variations in focal length as a function of wavelength for the objective of a single lens refractor, for two lenses (an achromat), and for three lenses (an apochromat). The steeper the curves, the more significant the chromatic aberration is in the spectral zone considered. Note that the typical spectral sensitivity of a CCD (from 500 nm to 1100 nm) falls in a less favorable zone than that of the eye or photographic film (around 550 nm).

FIGURE 3.4 An image of Messier 35 obtained at Bonsacre with a LYNXX camera equipped with a 135 mm F/D=3.5 photographic lens with a 1 minute exposure. The stars are surrounded by a fuzzy halo due to chromatic aberration of the lens, which particularly affects long wavelengths. The stars would have been less spread out if the image had been taken using a filter cutting out the near infrared. Notice in the upper right a large hazy spot which is the cluster NGC 2158. Documentation: Mathieu Sénégas.

Objectives described as being achromatic are composed of at least two different lenses composed of different glasses, and paired so that the focal length variation in relation to the wavelength is minimized for a certain spectral region. Unfortunately, this region is quite narrow. Commercial astronomical objectives are made so that this minimum dispersion region is centered on green, where the human eye is most sensitive; photographic lenses have the same property, since green also corresponds to the central chromatic sensitivity of photographic films.

CCD detectors have a wider spectral sensitivity than the human eye or photographic films, and are centered rather on red. CCDs, therefore, are very sensitive to a spectral region where the focal distance's dispersion in relation to the wavelength is relatively large for a standard lens or a photographic lens. As a consequence:

The image produced by a refractor and captured by a CCD suffers a certain chromatic aberration. This problem is not crippling and can be reduced with a Schott BG39 or KG3 filter, but it is preferable to mount a CCD behind a reflector whenever we have the choice.

3.1.4 *Guiding problems*

Long exposures require guiding: a star's image is observed with the help of an eyepiece equipped with cross-hairs; once the star moves from the cross-hairs' intersection, this signifies that the telescope is not following the sky's motion perfectly. The operator, therefore, corrects using the telescope's motors. The optical problem for guiding is to provide, on the one hand, an image that is captured by photographic film or by the CCD, and on the other hand, another image containing the guide star for the guiding eyepiece.

There exist three classical optical assemblies:

- Two independent telescopes are mounted in parallel on the same drive; one is equipped with a detector, the other with a guiding eyepiece (figures 3.5 and 3.6).

- At the telescope's focus, a semi-reflective thin plate, inclined at 45°, is placed in front of the detector and returns a fraction of the incident light to the side, thereby forming a second image of the same field, destined for the guide eyepiece (figure 3.7).

- At the telescope's focus, a small mirror, inclined by 45°, is placed in front of the detector, to the side of the field, and reflects the image of a small corner of the sky to one side of the object to be photographed. The hope

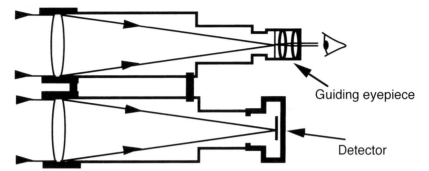

FIGURE 3.5 Two telescopes mounted in parallel, one with the detector, the other with the guide eyepiece.

FIGURE 3.6 Example of a 60/700 refractor mounted in parallel on a T180 *F*6 telescope used for guiding. Documentation: Vincent Letillois, SCCA Astronomical Club of Reims.

is to find a star useful for guiding in the little piece of the sky seen via the off-axis eyepiece (figure 3.8).

The two telescope assemblies and the semi-reflective thin plate assembly are just as useful in CCDs as they are in classical photography, since they do not require the camera to change position to find a guide star.

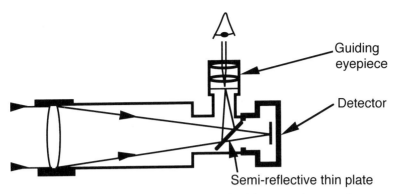

FIGURE 3.7 The guide eyepiece mounted on the side of the main telescope tube receives light from a semi-reflective plate inclined at 45°.

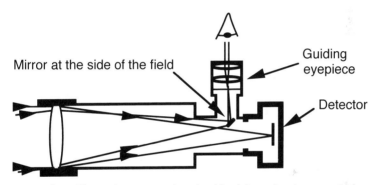

FIGURE 3.8 The guide eyepiece mounted on the side of the main telescope tube here receives light from a small mirror inclined by 45° is placed to the side of the field and reflects the image of just a small corner of the sky near the object to be photographed.

It is not as easy to use the system with a mirror at the side of the field. Telescope manufacturers suggest, among the accessories, a connection that allows one to affix a camera and contains a small mirror at 45°, with a lateral opening for the cross-hair eyepiece. The problem with such an assembly is the reduced size of the field accessible to the guiding eyepiece. To hope to find a guide star, the guide eyepiece must sweep the largest sky zone possible around the photographed object. With this goal, the mirror can generally move in and out toward the optical axis, in order to capture areas of the sky that are more or less close to the center of the photographic field; also, the connection can turn around the telescope's optical axis so that the observed zone moves around the photographed object. The problem is that the connection carries the photographic camera (or the CCD camera) in its rotational movement.

In classical photography, this camera rotation is not inhibiting. But it constitutes a major problem for CCDs because of the need for flat fields (as we shall see in section 3.4): in section 3.6, we will see that the camera's position must not be changed between the flat field acquisition and the image acquisitions. Normally, the flat fields are acquired on the twilight sky. It is therefore not possible to turn the camera on its axis between each image. Also, the more or less deep position of the mirror can modify the light flux that reaches the CCD detector and hence seriously modify the flat fields at the edge of the field.

Classic guiding systems, which implies searching for the guide star by rotating the housing, are therefore incompatible with CCD images. Also, Astro-Equipment, a company specializing in CCD cameras, have produced a special guider for CCD cameras: the connection that attaches the camera body onto the telescope contains a flat mirror at 45° with a hole pierced in its center: the light destined for the CCD detector, in the center of the field, passes through the hole of the mirror, while light from the side of the field is turned to the side by the mirror to the guiding eyepiece. The latter is mobile and can examine the entire periphery of the field by looking at the mirror zone in front of it. The search for a guide star, therefore, can be done while leaving the camera body fixed.

3.1.5 *The filter assembly*

As in photography, the use of colored filters allows the taking of pictures in precise spectral regions. It is even commoner to use filters in CCDs than in photography, although for different reasons:

- A CCD's sensitivity being superior to a film's, it is possible to use a filter with a CCD in circumstances where a photographic filter would lead to prohibitive exposure times.

- For a refractor, the use of a narrow band filter can considerably reduce chromatic aberration.

- Whereas the three-color process is a difficult photographic exercise, it is particularly easy to accomplish with digital images, and thus it is tempting to create images in the three primary colors to obtain a composite image in true color.

The filters are placed in the converging beam, slightly in front of the detector. In photography, we often use filters that are fixed on or screwed onto the camera. To change filters, the camera is unmounted from the telescope, the filter is replaced, and then reassembled.

Unfortunately, for CCDs, the camera body must not be turned during the night's observation in order to be able to use the flat field acquired at sunset.

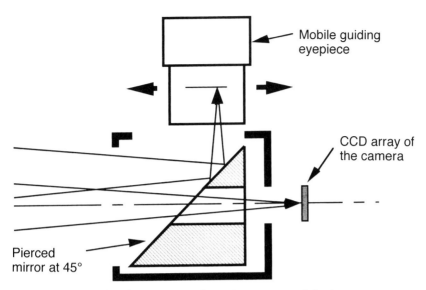

Mobile guiding
eyepiece

CCD array of
the camera

Pierced
mirror at 45°

FIGURE 3.9 A special guiding connection for a CCD camera, made by Astro-Equipements: the guide eyepiece is mobile horizontal above the 45° mirror to examine the field's periphery; the light from the center of the field reaches the CCD through the hole pierced in the center of the mirror.

This has important consequences for the way the filters are mounted. Not only must the mechanical assembly allow the changing of filters without moving the camera body, but each filter must be placed in a perfectly reproducible position: indeed, a flat field will have to be produced with each filter, which will keep track of the filter's transmission defects; each of these defects (dust, dirt, inhomogeneities, etc) must be placed in front of the same CCD pixels during the flat fields and during the imaging session, a few hours later; therefore, in the meantime the filter will have been removed and replaced. All the filters that will be used during the night must be placed in the filter holder during the day in order to be available to produce the flat field once night falls. There exist two groups of filter carriers: the wheel or the sliding drawer. No matter what system is used, one must be able to install at least 3 filters (BVR). It is imperative that the positioning of the filters be reproducible in order to preserve the defect positions that will be corrected by the flat field. For this, a device which positively locates each filter is used. The wheel or drawer must click at each position to be sure that we always return to the same position.

Some filter wheels are controlled by stepping motors. The positioning of a filter corresponds to the rotation of a certain number of motor steps. The reproduction of position, therefore, is not as good as that of a wheel with

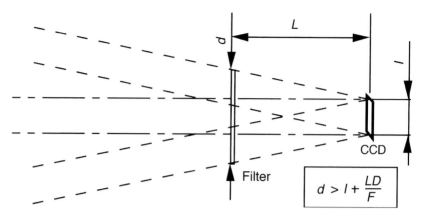

FIGURE 3.10 This diagram shows that one must select the filter size larger than the diameter of the full light beam (without vignetting). l is the length of the CCD's side, L is the distance separating the CCD detector from the filter, and F and D are, respectively, the diameter and the focal length of the telescope.

mechanical reference points and can be inadequate in terms of the CCD's requirements.

A good precaution consists in protecting the filters from dust and not moving them too violently during the observation session: just one speck of dust between the flat field instant and that of the image could translate into a small defect on the image. In astronomy, there are standard color systems (see section 4.4.1) and narrow band filters centered on specific wavelengths are also used, allowing the isolation of structures of one type or another in the images (dust or gas for comets, emission lines of planetary nebulas, molecular bands for planets, etc.). It is essential to use a combination of glass and not gelatin filters since the latter transmits the near infrared, a spectral region in which CCD cameras are sensitive while photographic films are not. Hence, a gelatin filter, designed for photography, could have an unexpected effect in CCD imaging. The polarimetric study of celestial objects requires the use of 3–4 polarizing filters. Just simple, inexpensive filters allow good observations. Unfortunately, polarimetric analysis often requires the use of two filters at the same time: a colored filter to select a wavelength range and a series of polarizing filters to measure the amount and angle of polarization. Therefore, two independent filter holders are needed! During the positioning of the polarizing filter in the filter holder, the polarizing angle adjustment must be done with precision.

Finally, one must never forget to calculate the diameter of the optical beam at the level of the filters and lenses, in order for them to be chosen large enough that they do not act as a diaphragm and cut out the periphery of the beam.

The use of filters in CCD astronomy is very common. But the mechanical filter positioning system must allow the successive use of several filters in the same night while respecting the following two conditions:

- The camera must not be unmounted and must spend the entire night in exactly the same position it was in during the flat field acquisition.
- Each filter must return to the exact position, for each image, it held during the flat field acquisition.

3.1.6 *The mounting for TDI mode*

We have seen that the TDI mode consists in letting the sky pass before the camera and read the image at a speed strictly equal to that of the sky. It is therefore necessary that the direction of the CCD's columns, that is, the direction in which the charges are transferred, are strictly parallel to the east–west direction of the sky image's natural motion.

Consider, for example, a Thomson 7895 CCD with 512×512 pixels. If the CCD is not perfectly oriented, the captured image in TDI mode is blurred in the direction of its declination. This blur reaches a value of 1 pixel if the array's orientation is incorrect by only $1/512$ of a radian, or around $0.1°$. Such precision cannot be reached by simple friction positioning. It is necessary to produce a small mechanism that allows the camera head to rotate around its optical axis, with a micrometric screw adjustment. TDI mode allows the production of images while the telescope remains pointing in a fixed direction. The temptation is great, therefore, to be content with an azimuthal mounting, or even consider using images from a Dobsonian telescope, for example. If the mounting is equatorial (and correctly set up!), the direction of the sky's motion at the instrument's focus does not depend on the targeted sky area. But if the mounting is azimuthal, the direction varies in accordance with the targeted point.

If we hope to produce images in TDI mode indistinguishable from conventional images with an azimuthal mounting, we must therefore adjust the camera body in the right direction for each image, which quickly risks becoming tedious and time consuming. The best solution, therefore, is to adjust the azimuthal mounting to the level of the meridian and produce images of objects as they cross the meridian: the apparent direction of the sky's motion being constant along the meridian, all of the observable objects are thus accessible at the cost of only one adjustment.

Images in TDI mode require the use of a mechanical micrometric mounting, which allows the camera to be oriented around the telescope's optical axis with a precision superior to a tenth of a degree.

3.1.7 *The first images*

You have just removed the camera from its packaging. How do you quickly verify if it is in working order? Firstly, read the documentation and carefully follow the connection procedures. Secondly, before installing the camera on the telescope, we suggest three tests.

We begin, for the first test, with the cooling. Wait 10 minutes to be certain that the camera's temperature is stabilized. Make a 10 second exposure with the aperture pointed to an intense light source (a 100 watt light bulb, for example). You should obtain a completely saturated image (while on the screen). If all of the camera's pixels are at zero (black image on the screen), check the feeds and connections to the computer.

The second test concerns the precharge image. For this, the camera's aperture must be efficiently blocked from light (using several black rags, for example) and the shortest possible exposure must be taken (generally 0.01 seconds). Using the cursor cross-hairs, verify that the pixel's values are above zero by a few percent. If this is not the case, start worrying. Save this image on the hard disk.

The third test concerns the dark image. For this, in the same conditions as for the precharge, take a 60 second exposure. Then, subtract the precharge image previously taken from the dark image. The average level of this image will give you the number of ADUs per minute due to thermal loads. If the camera is saturated in 60 seconds, verify if the Peltier coolers are well connected.

These three tests allow one to see if the camera is well cooled and that it responds well to light. We can then proceed to mount the camera onto the telescope. On the first night, target the Moon to make adjustments in automatic mode. Also, take advantage of the evening by centering the telescope lens in relation to the CCD's field. The rest of the night should be devoted to tracking and targeting practice on stellar fields or deep-sky objects.

3.2 Pointing the telescope

The inconvenience of keeping the CCD camera in a strictly identical position from the flat field acquisition to the image acquisition forbids removing the camera body from the eyepiece holder to replace it with an eyepiece when an object in the field must be targeted. Once the camera is replaced to acquire the object's image, one is no longer guaranteed that it will be in exactly the same position it had at the moment of the flat field but maybe within a fraction of a pixel, or a few micrometers. We can contemplate the use of an extremely precise mechanical positioning system, but this solution is not used because of its high production cost and limited interest.

The telescope targeting, therefore, is done with the camera in place. It is the camera that provides the necessary images. The operator no longer has an eye on the eyepiece but is seated in front of a computer screen: there is a gain in comfort, but the computer must be close to the telescope.

Targeting done directly on the CCD camera images has an advantage as well as a difficulty compared to examining the field with the help of an eyepiece. The advantage is linked to the CCD's great sensitivity. With an integration time of a few seconds and quick processing we see objects appear which the eye is incapable of perceiving in the eyepiece. This is an important advantage when searching for faint objects: for example, a 5 second exposure through a 20 cm telescope allows the identification of a 13th magnitude galaxy.

The difficulty is the CCD detector's small size, and hence the small field covered. A LYNXX camera at the focus of a 2 meter focal length telescope covers a field of only 4.5'. This field is sufficient to contain most faint galaxies accessible to this instrument, but finding a coveted galaxy could present some problems.

Luckily, several solutions exist:

- Several telescopes are equipped with precise setting circles so that targeting precision is more accurate than the size of the field .
- A small refractor and a large finder, in parallel, guarantee most targetings; but the original finders provided with commercial instruments (an 8×30, for example) are often inadequate.
- An adaptor between the CCD camera and the telescope can be made which contains either a thin semi-reflective plate or a removeable mirror inclined at 45° to the optical axis in order to return the targeted field's light toward a guiding eyepiece.
- The guiding probe on the edge of the field from Astro-Equipment described in section 3.1.4 can also be used for centering since it allows a view of the area around the zone targeted by the CCD camera up to 20 mm from the center, or around 1° with a 2 meter focal length telescope.

The trend is toward larger and larger CCDs, in amateur cameras. For example, the Alpha 500 camera allows a field of 15' per side to be covered, or half of the lunar diameter, at the focus of a 2 meter focal length telescope. It is therefore possible to target without the help of special equipment.

Targeting can be tricky with a small CCD covering a reduced field. It is therefore useful to plan for a system that uses targeting aids (a guiding refractor, coordinate circles, etc.).

FIGURE 3.11 A CCD camera mounting equipped with an adaptor enclosing a semi-reflective thin plate returning 20% of the light toward a guide eyepiece (vertical in the photograph). The telescope's connection tube contains a Barlow lens allowing the focal length to be increased. The special eyepiece holder enables focusing without any field rotation. Documentation: Gino Farroni.

3.3 Focusing

It is imperative to accurately focus the image before beginning the night's observations. The goal is to place the CCD's sensitive surface co-incident with the telescope's focal plane.

The ban on unmounting the CCD camera body between the flat field and the observation is a great handicap. It is therefore impossible to do a Forcault test (seeking the focus by cutting with a knife edge a light beam coming from a star and focused by the objective), which is the most efficient method of focusing in astrophotography. Then again, we can consider using a mechanical connection for affixing the CCD camera to the telescope, which would contain a removeable mirror at 45°, directing the light to the side toward a Foucault knife edge. The biggest difficulty, therefore, would be to guarantee that the optical distances between the telescope's objective and on the one hand, the CCD's sensitive surface, and on the other, the Foucault knife edge are exactly the same. The best solution to reach this equality would be to provide an adjustment system for the knife's position and focus on the star simultaneously through a Foucault test and the CCD image. The adjustment precision of the instrument is thus limited by the focus quality on the CCD image. In photography, the equality of the optical distances is guaranteed by directly applying the knife edge in place of the plate holder, which eliminates by definition all mechanical imprecision and makes this a very effective method.

If we eliminate all Foucault test possibilities, we are only left with focusing by directly judging the quality of the image delivered by the CCD. Luckily, we shall see that electronic imaging through its powerful display options, makes it easier to judge the focus quality than through a photographic camera's view finder.

To be efficient, focusing must be done on the image of a star.

We should choose a star bright enough to provide a large signal (at least half of the numerical scale, while checking that it does not saturate) with short exposure times (for example, a tenth of a second) in order to freeze the turbulence. During the focusing, the camera is used in an automatic image mode windowed on the star. The focusing device is controlled in such a way that the star's image appears as point-like as possible.

Note that the integration time must be shortened as the correct focus is reached: indeed, the star's light becomes more and more concentrated on a small number of pixels, which can saturate; of course, it is not possible to judge the focus if a few pixels from the stellar image are in saturation.

It is necessary to wait a while between each correction, so that the image displayed on the screen is actually the one produced by the camera after the last correction and not the one before! Otherwise, it is suggested that one view several images for each correction in order to compensate for the variations in seeing due to atmospheric turbulence.

Several image display functions can help the operator.

Firstly, the 'magnify' function allows the enlargement of a star's image until one can count the number of pixels it covers. However, beware the fact that the number of pixels covered also depends on the choice of the low visualization threshold: if the chosen threshold is too high, only the most illuminated central pixels will appear white on the screen, which does not mean that the star does not extend over many other pixels all around with a lower illumination than the chosen threshold.

A good way of avoiding this is to choose a stellar image display in false colors:

the human eye can thus visualize a wide range of illumination. As each color represents a star's isophote, it is easy to follow the display by taking into account the boundary of the same color from one image to another. Incidentally, the appearance at the center of the stellar image of a color corresponding to increased illumination signifies a larger concentration of light, and hence a better focus.

A particularly efficient aid is the photometric profile: the acquisition software traces, for each image, the light intensity variation from one pixel to another along a line passing through the most illuminated pixel, that is, through the center of the stellar image. The proper focus is reached when the image profile, thus obtained, is the narrowest possible. One way to quantify the thinness of this profile consists in measuring its full width at half maximum, which constitutes an objective estimate of the quality of the focusing. High performance acquisition software, such as that of the Alpha 500 camera, has a magnifier, a photometric profile on two perpendicular axes, and a half maximum measurer. Temporal fluctuations of the image quality, displayed by the means just described, otherwise allow a good evaluation of atmospheric turbulence.

With amateur telescopes, focus can vary over the course of the night owing to expansion or contraction of the telescope or deformation of the optics with temperature. Never hesitate to refocus several times a night.

Let us remember a fundamental point already addressed in section 3.1.4: to focus with a CCD camera we must never use a screw eyepiece holder whose tube turns during the adjustment. For if the camera turns, the CCD field also turns and the image produced no longer corresponds to the flat field.

Note that it is much easier to focus when the telescope has a large F/D ratio. In fact, the theoretical focal plane, allowing us to reach the diffraction spot, is equal to $7.08 \times \lambda \times (F/D)^2$. For $\lambda = 0.65$ µm, with an instrument at $F/D = 3.5$, the accuracy of the focusing must be 56 µm, that is, 0.056 mm. With an instrument at $F/D = 10$, the accuracy of the focusing must be 460 µm, that is, 0.46 mm.

In practice, for various optical aberration reasons, the range of the focal plane often seems more extended (by a factor of at least two). When one uses filters, the focal plane retreats by about one third the thickness of the filter. When one wishes to do trichromatisms on planets, one should try to use filter combinations of the same thickness for each color in order to avoid refocusing between each color. In addition to focusing, the shape of an unfocused star's image allows us to diagnose some optical problems. With a reflecting telescope, once it is very unfocused, a star's image appears as a large ring of light. In fact, the interior black disk is the secondary mirror's shadow. When the CCD is inside

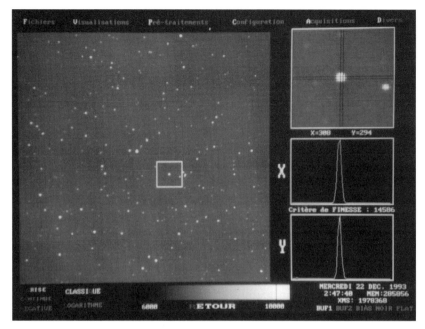

FIGURE 3.12 A screen showing the image acquisition software from the Alpha 500 camera. The software offers several complementary ways to control the focusing of a star. Documentation: Mathieu Sénégas.

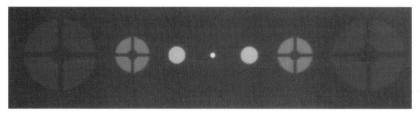

FIGURE 3.13 The evolution of a star's image over the course of a focusing sequence. When the star is very unfocused, either intra-focally (left) or extra-focally (right), the shadows of the secondary mirror and the spider can be seen. The star is focused when its image is as small as possible (center).

the focal plane, this is the intra-focal region. In the opposite case, it is the extra-focal region. The shape analysis of an unfocused stellar image allows the diagnosis of some optical problems.

If the black disk is off center with respect to the light disk, this is due to comatic aberration. This also means that the star whose image is being produced is not on the mirror's principle optical axis. The optical axis is located

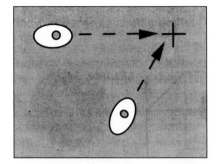

FIGURE 3.14 In the presence of coma optical aberration, the unfocused image of stars outside the optical axis show a radial extension in the direction toward the optical center symbolized by a cross.

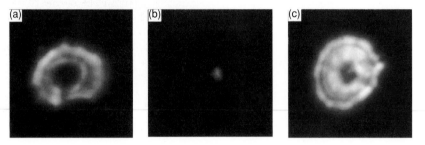

FIGURE 3.15 Focusing sequence performed with a telescope opened at $F/D=14$, showing astigmatism. (a) The intra-focal image (-3.3 mm) shows a horizontally extended light ring. (b) The focused star. (c) the extra-focal image ($+3.3$ mm) shows that the star's elongation has turned through $90°$.

somewhere along the elongated out of focus stellar image. With several stars in the field, we can find the exact position of the optical center. Hence, we can easily center the camera on the optical axis. Remember that the center adjustment step is just as important as the focusing step.

Other telescopic optical aberrations can also come to light. If the intra-focal unfocused image is extended into an ellipse, this signifies that the optics suffer from astigmatism. We can check if this is the case by noticing if the image becomes round when one achieves the best focus and then returns to an elliptic shape when one moves to the extra-focal region (the ellipse's major axis must have turned a quarter turn in relation to that of the intra-focal ellipse).

Once the CCD detector is well centered on the optical axis, sometimes the presence of a light ring in the intra-focal position can be noticed. This is a problem due to spherical aberration. This was the major optical problem of the *Hubble Space Telescope*. Once we move to the best focus point, the ring decreases in size and disappears to become a bright spot. It is just prior to this position that we find the best focus. Then, the extra-focal image shows a central peak surrounded by diffuse light without a ring.

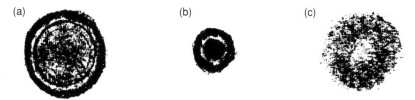

FIGURE 3.16 A schematic of a focusing sequence showing the effect of spherical aberration. (a) The intra-focal image shows concentric rings of light, of which the brightest is the outermost one. (b) The focused star. (c) The extra-focal image shows a pale diffusion. The images are shown in negative.

Finally, if the outside of an unfocused stellar image seems 'broken' or raised, the secondary mirror is held tightly and has mechanical stresses.

3.4 Photometric correction of images

Figure 3.17 shows the different intensity contributions $I(x,y)$ that are measured on the coordinate pixel (x,y) of the raw image which are stored in the computer's memory. The preprocessing consists in extracting the $i(x,y)$ value, which represents the luminous intensity that actually impinges on the pixel (x,y). The different contributions to intensity $I(x,y)$, on the raw image, are defined by three fundamental parameters:

$b(x,y)$: the bias is a precharge value of the CCD, constant, and independent of the temperature and exposure time.

$d(x,y,t,T)$: the dark is the value of accumulated thermal loads during the exposure. The value of $d(x,y,t,T)$ is proportional to the exposure time t. For the same exposure time, the value of $d(x,y,t,T)$ is as small as the temperature T is low.

$r(x,y)$: the response factor of the pixel. It is determined from the flat field image.

The relation that links $I(x,y)$ to the parameters defined below is as follows:

$$I(x,y)=b(x,y)+d(x,y,t,T)+i(x,y)r(x,y).$$

Since we are looking to extract the $i(x,y)$ quantity from the raw image $I(x,y)$, the above relation becomes

$$i(x,y)=\frac{[I(x,y)-\{b(x,y)+d(x,y,t,T)\}]}{r(x,y)}.$$

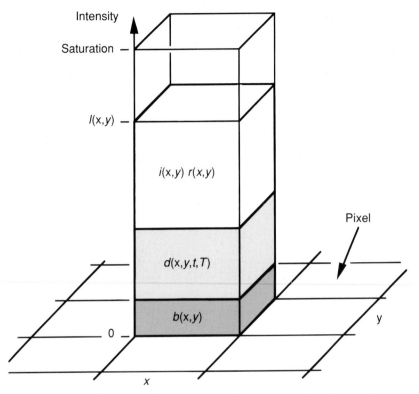

FIGURE 3.17 The different contributions to $I(x,y)$ that are measured on the coordinate pixel (x,y) of the raw image which are stored in the computer's memory.

In practice, the $b(x,y)$, $d(x,y,t,T)$, and $r(x,y)$ values must be linked to calibration images. Since there are three values, three calibration images are needed. They are as follows:

> The bias: this is an exposure, the shortest possible, in total darkness:
> $$I_b(x,y)=b(x,y).$$
> The dark: this is an exposure, taken in total darkness, with the same time t at the same temperature T as the image to be preprocessed.
> $$I_d(x,y)=b(x,y)+d(x,y,t,T).$$
> This exposure contains the precharge.
>
> The flat field: this is an exposure with a length of t_f, done on a flat field (a blue sky, for example). $i(x,y)$ is therefore constant for all of the array's pixels.
> $$I_f(x,y)=b(x,y)+d(x,y,t,T)+r(x,y)\times\text{constant}.$$
> In most cases, the flat field's exposure time is sufficiently short (t_f being

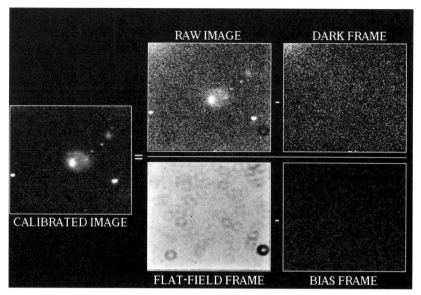

FIGURE 3.18 A global view of different images that intervene during the course of a raw image's preprocessing stage. In this figure we have placed the images as in the $i(x,y)$ equation shown below. The image is of NGC 2793: Patrick Martinez, Association T60.

less than 5 seconds) to ignore the thermal charge contribution, and the equation then becomes

$$l_f(x,y) = b(x,y) + r(x,y) \times \text{constant}.$$

The equation linking $i(x,y)$ to $l(x,y)$, previously established, therefore becomes

$$i(x,y) = \frac{[l(x,y) - l_d(x,y)]}{[l_f(x,y) - l_b(x,y)]} \times \text{constant},$$

where the constant is valued as the average of all the flat field's pixels subtracted from the precharge in order to keep the initial dynamic range of the raw image.

In practice, the preprocessing phase, or photometric correction, consists firstly, in synthesizing a new image from a raw image from which the dark was subtracted, and secondly, in taking this new image, which we divide by the flat field (itself previously subtracted from precharge) and multiplied by a constant.

3.5 Dark acquisition

For the image's pretreatment phase, we have seen in the previous section that the first operation consists in removing the dark current's contribution from each pixel. We must therefore measure, at a given moment, the dark current's value. This is done by producing an image in absolute darkness, by leaving the camera shutter closed, for example. This image, whose pixels contain only the dark current's and the precharge's contributions, is called the dark image, or simply the 'dark'.

The most concise image treatment software simply allows the subtraction of two images, pixel by pixel. We therefore remove a dark image from the sky image. Thus the dark image must contain a dark current exactly identical to the sky image. Therefore, it must be produced with an identical exposure time and temperature. If the CCD temperature is not regulated, which is the case for many amateur cameras, it is then necessary to obtain the dark image just prior to or just after the sky image, since ambient temperature variations during the night, or a different ventilation of the camera body induces CCD temperature variations. In practice, the observer must produce a dark of x seconds just after the sky image of x seconds. In fact, we are often content to acquire a single dark image for a series of sky images with the same exposure time and limited in time. The systematic acquisition of a dark image after each image is not as crucial a problem as it appears: CCD exposures are generally quite short, 5 minutes, for example: once the exposure is finished, we can ask the camera to take a 5 minute dark image, which corresponds to the time necessary to point the telescope toward the next object; therefore, there is no lost time.

Advanced image processing software has a function which adjusts the level of any dark image to the dark signal of the raw images (see section 5.2.2). Thus, it is possible to use a dark image, produced with whatever exposure time, with a different CCD temperature. This offers two advantages in terms of the previous method:

- It is no longer necessary to produce dark images during the observation session: one would always use the reference dark image acquired just once;

- The reference dark image may have a completely different dark level to subtract from the raw image; we would therefore use an image close to saturation, obtained by an exposure or a series of exposures accumulated for a total value of at least half an hour.

Since relative thermal noise is inversely proportional to the square root of the number of electrons accumulated, a dark image near saturation represents a very weak relative noise: after processing, the sky's image will represent thermal noise created solely by its own thermal charges, without the dark image used contributing in a significant way, while the image treated by simple subtraction will suffer from its own thermal noise and that of the dark image, or in the end a noise signal twice as large.

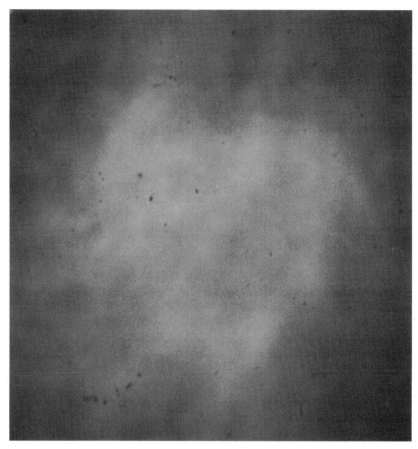

FIGURE 3.19 The image of a flat field obtained with an Alpha 500 camera (512×512 pixels) behind a 500 mm focal. Notice the vignetting effect (center lighter than the sides) and the presence of dust on the CCD (clean dark spots).

3.6 Flat field acquisition

In the image's pretreatment phase, we have seen in section 3.4 that the second operation consists in dividing the image by a flat field. We must therefore acquire a flat field.

The flat field takes into account the gain irregularities of the array's different pixels and the optical transmission irregularities, in particular, vignetting and the presence of dust in the light path. We must therefore produce a flat field for each optical assembly and, if possible, at each observation session, since dust can appear, disappear, or move from one day to another.

The main constraint comes from the variation in illumination memorized by the pixels, which corresponds to the precise locations of the pixels at the moment of the flat field acquisition. It is essential that the same pixels be at the same place during sky image acquisition. For example, if dust is found in front of pixel A at flat field acquisition (the pixel, therefore, registers a weakened luminous flux) and in front of pixel B during sky image acquisition, because of a displacement of only a few micrometers, the image processing will consequently not be able to re-increase the light from pixel B (which actually received less flux during the sky image), and will wrongly increase the signal received by pixel A (since we expected it to receive a fainter light). We can therefore create by accident, artificially darkened zones and artificially luminous zones during the image processing. To avoid this, it is imperative not to move the camera body between the flat field and the sky image acquisitions, which imposes some important operational constraints.

The other major flat field problem is to find an object which has a strictly uniform luminosity. The ideal solution consists in capturing an image of the sky at twilight, preferably in the opposite direction to the sunset, since the light gradient is weaker. The sky's luminosity is, in fact, sufficiently uniform for the scale of field covered by the CCD. But during the day, this luminosity is too strong and causes CCD saturation, even for short exposure times. However, one must avoid taking the flat field too late, once night begins to fall, because the integration time would be too long and the CCD risks detecting stars in the field, which would not constitute a uniform field. If the exposure's duration is defined by the opening of a mechanical shutter, it is necessary to use, for the flat field, a relatively long integration time in relation to the shutter operating time. Otherwise, obvious lighting differences can appear from one point on the CCD to another because of the shutter curtain's travel time. For the same reason, the flat field integration time must be long in relation to the charges' transfer time when the shutter is electronic on a frame transfer CCD. Because of these considerations, the optimal integration time for a flat field is in the order of a few tenths to a few seconds, which corresponds to the sky's luminosity a few minutes after sunset. During the flat field acquisition, the telescope is fixed or tracking at sidereal rate.

It is clear that the production of flat fields on a twilight sky entails several constraints:

- It is necessary for the system to be operational, with the CCD mounted and cooled, at the moment the sun sets, although the first images will not be possible until an hour later.

- The optimum moment for flat field production only lasts a few minutes, and at least one flat field must be done per filter used during the night, because each filter has dust and density irregularities which are unique to it; it is even preferable to take several flat fields with each filter in order to average them so as to reduce the influence of read noise and of a possible star (in fact, one synthesizes a median flat field, described later). Note that it is trickier to produce flat fields when the length of twilight is short (in winter, for example).

- Since it is imperative not to change the optical configuration between the flat field and the image acquisitions, we can work with a maximum of two optical configurations per night: one at the night's start, for which we produce the flat fields in the evening, and the other at the end of the night, for which we produce the morning flat fields before sunrise (which requires some self-sacrifice).

- The flat field must be produced after focusing, in order for the light path to be the same during the flat field and the nocturnal sky image. Thus, the focusing is done on the stars, and the flat field must be finished before the first stars appear! The solution to this paradox consists in keeping the camera mounted from one night to another, or to accurately find the position that corresponds to the focus. In all cases, the first night's flat field, produced on an approximate adjustment, will not necessarily be very satisfying.

In case the twilight flat field cannot be produced, a helpful solution consists in making an image of a screen lit in the most uniform way possible and placed just before the telescope's opening. One must not be deceived by the uniformity of a screen's artificial lighting: variations, imperceptible to the human eye, will not go unnoticed by the CCD, which is capable of revealing irregularities weaker than one in a thousand. As the telescope obviously is set to infinity, the screen's image is totally unfocused but has the drawback of lighting the CCD with light different than that produced by a nocturnal sky.

The flat fields' integration times must be adjusted so that the pixels accumulate a number of charges near full capacity. Hence, the read noise and the photon noise have a relatively weak influence; we can further diminish this influence by averaging several flat fields. Of course, it is fundamental that no pixels reach saturation; also, one should avoid overfilling the pixels if they show a non-linear response close to saturation (one should fill, on average, between a half and two thirds of the dynamic range). After each flat field acquisition, it is prudent to display the image's histogram in order to be certain of the pixels' proper filling.

The LYNXX cameras have a trap when it comes to this: a special function performs the flat field acquisition and automatically processes the image obtained in a way that the final histogram is centered on the average capacity value; hence, even if the integration time is too large and the pixels saturate, the flat field histogram has a reassuring appearance but is misleading because it is

FIGURE 3.20 A plate covered with light diffusing material is fixed onto the interior of the dome in order to produce the flat fields because they cannot be produced from the sky. Documentation: Gino Farroni.

centered in the middle of the scale. The experienced operator will notice that this histogram is, nevertheless, abnormally narrow (all the pixels have the same value, equal to the saturation value divided by 2), while the accumulated charges in the pixels during a flat field normally represent a disperson of a few percent. The correct solution with a LYNXX camera consists in defining the correct exposure time by the normal image acquisition function followed by the display of a histogram, then applying the exposure time to the flat field acquisition function. A technique often used by professional astronomers allows the synthesis of a flat field from night sky images. A flat field obtained in this way is called a 'median flat field'. Consider an odd number of images produced in the same conditions, the same night, say $2N+1$ images. After having removed the thermal charges, each pixel from each image contains a charge created by a star, or light interference from the sky (if no star falls on the pixel of the image). Stars occupy a small area on a deep-sky image, so this last case will be the most common. On all of the chosen images, each pixel has received $2N+1$ different values (one in each image). The median flat field is created by taking for each

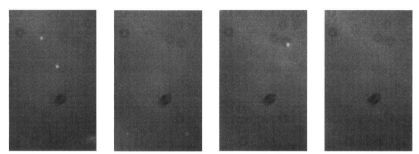

FIGURE 3.21 The three images to the left are individual flat fields produced from the sky and containing star images. The image to the right is the median flat field of the three preceding ones. It no longer shows a star.

pixel the median value of its $2N+1$ values; the median value is defined as the value such that N values of the whole will be larger and N values smaller.

However, it is necessary to have normalized all of the individual flat fields on the same average. On condition of not having systematically centered an object on the same field zone, the center, for example, of each pixel would only have been touched by a star in far fewer than half of all cases, and since these cases give charge values greater than the sky brightness, they are eliminated by the median filter. The median flat field, therefore, is an image constituted of several pixels uniformly lit by a uniform sky background. The trouble with this is its low level of illumination, in comparison with which the read noise and the thermal noise cannot be negligible. Luckily, the median filter considerably reduces these noises; meanwhile, the median flat field method, which is an efficient and much less restrictive method than the dusk sky flat field, is only recommended for very good cameras that have a low read noise and a very weak dark current.

3.7 A night of observations

An hour before sunset, we begin by starting the CCD camera coolers. The computer's internal clock must be checked to see if it indicates the same time as a standard time signal. Indeed, some CCD image acquisition software causes errors in the computer's internal clock.

After ten minutes of functioning, we can produce the dark and precharge images in order to control the thermal and read noise values. Use the time to check the detector's temperature and to identify the possible presence of electrical interference. If there is time, and if it has not already been done, we could synthesize the generic dark image as described in section 5.2.2. Starting from the moment the Sun sets, point to a bright star or the Moon to roughly focus, if this has not already been done.

We must now watch for the best moment to make the flat fields. We suggest placing the camera in automatic mode, taking a 5 second exposure, displaying the images with a low threshold equal to zero and a high threshold at the dynamic range's maximum (4096 for a 12 bit camera). At first, there will still be too much daylight and the images will be saturated (white on the screen). As the day fades, there is a moment where the images will no longer saturate. We must guarantee, with the help of a statistical analysis function, that not a single pixel is saturated. We can then begin saving a dozen flat fields onto the disk. If colored filters will be used during the night, a flat field must be done with each filter. Since the camera is not very sensitive to blue, we should begin the flat fields with the B filter. Then, in order, we use the V, I, and red filters for the flat fields.

Once the flat fields are done, the night sky is not completely black. We should take advantage of this and synthesize the averaged flat fields or medians by the required method (see section 5.2.3). Then, we recommend keeping the camera pointed at the same object as at the beginning of the night to refine the focus. We can also judge the image quality by comparing it with images from previous nights. We should also take the opportunity to do a complete prepro-cessing of this image in order to judge the synthesized flat field quality. If the preprocessing is satisfactory, we can subtract out the raw flat field images which were used to synthesize the averaged or median flat fields.

Over the course of the night, once a new image has arrived, its quality must be assessed, focus (star image display), saturation (locate the saturated pixels), motion (the telescope's tracking), interference (periodic structure in terms of the sky background) and cold control (the sky background's average value for a given exposure time). For this, we suggest that the raw image is visualized with high and low thresholds on both sides of the sky background's value.

It is highly recommended to keep a notebook or paper to keep details of the file name which is saved onto disk, the observation time, the targeted object, and the filters used. This may appear redundant since the computer 'theoreti-cally' registers the date and time automatically in the file, but experience shows that the notebook allows one to find the cause of an image problem more easily.

If one has the courage or stamina to continue observations until morning, one should not forget to make new flat field acquisitions at dawn.

Some CCD image acquisition systems provide corrective options, in real time, of raw images arriving to the camera. They consist of a preprocessing exe-cuted with the classical methods described in section 5.2.2. The hard to please user would prefer, however, to carry out the preprocessing differently by one of the methods using generic darks.

4 Display and image analysis functions

4.1 The concept of the CCD image

In the preceding chapters, we have detailed the functioning of the CCD camera. The following chapters will describe the different ways in which the images from the camera can be manipulated. But first, what do we mean by CCD image?

The camera is placed at the focus of a telescope and allows an electronic image to be generated; we will call it the 'image on the CCD detector'. This image is recovered by the computer which stores it in its memory: this is the 'image in memory'. We can then display it on a screen: this is the 'image on screen', or, once again, store it on a magnetic disk: this is the 'image on disk' (see figure 4.1). It is necessary to master these different image forms before approaching the display techniques and image processing which is why we will now give more attention to them.

4.1.1 *The image on the CCD detector*

We know that the CCD camera is made up of a pixel array. The image is projected onto the CCD detector, therefore, is broken down into tiny squares called

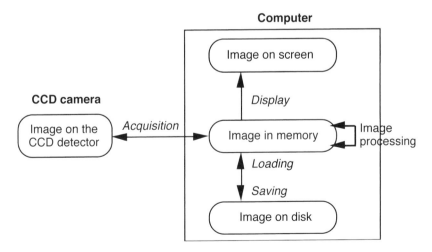

FIGURE 4.1 A synoptic schematic of the idea behind a CCD image.

10	10	10	10	10	10
10	10	200	10	10	10
10	10	10	10	10	10

FIGURE 4.2 An array representation of an image in memory. Each array square represents the pixel that corresponds to the image formed on the CCD detector array.

pixels; this is the image sampling. We have seen that each of the CCD detector's pixels converts received light into electrons. The image on the CCD detector, therefore, can be compared to an array in which each pixel contains a certain number of electrons.

The image on the CCD detector is not of great interest for the camera user, since the goal is to have a visual rendering of the image. To achieve this, we must transfer the contents of the CCD detector's image to the computer's memory. It is during the course of this transfer that the electrons contained within each pixel will be transformed into ADUs (analog digital units) by the analog–digital converter. Thus, each image pixel is given a number of ADUs proportional to the light received by the corresponding pixel on the CCD detector. We will see, in section 6.4.1, how ADUs are linked to a magnitude unit.

4.1.2 *The image in memory*

The image that exits the CCD camera arrives in the computer's memory. The RAM (random access memory) and the image are erased as soon as the computer is switched off. The image, therefore, can be compared to a dream; it ends as soon as we wake up.

The memory image is represented by an array. Each pixel from the array contains a number of ADUs assigned during the CCD detector's transfer to the computer.

For example, the array of an image in memory is 6 pixels by 3. A star occupies a single pixel, at coordinates (3,2) and generates 190 ADUs. The remainder of the sky uniformly generates 10 ADUs on all of the array's pixels. We can therefore use the representation mode in figure 4.2.

Beware: in all CCD systems there is a maximum value that the ADUs can never exceed which depends on the digital converter. As we have seen, converters usually digitize on 8, 12, or 16 bits, which respectively correspond to the maximum values of 255, 4095 and 65535 ADUs.

4.1.3 *The screen image*

The on screen image is a colored representation of the numbers contained in the memory image's pixels. The color principle is very simple: the software is equipped with transcoding tables, often called LUTs (look up tables). A LUT represents the color associated with each ADU value.

The most natural display of the image is the gray level representation. The natural transcoding table is constituted in the following way: the colour black is associated with pixels at 0 ADUs; white is associated with pixels with 4095 ADUs (in the case of a 12 bit digitized image) and for the pixels whose value lies between 0 and 4095 ADUs, the computer generates different gray levels from very dark to very light. The natural transcoding table, as described above, runs into some difficulties: in the example of the 190 ADU star on a sky background of 10 ADUs shown in figure 4.2, the display of the image on the screen will show a gray star on an almost black background. That being said, it will be difficult to properly see the contrast between the star and the sky background. Figure 4.3 shows the image screen in this case.

There are two ways to artificially increase the contrast:

- First method. We exposed the values contained in the memory image's pixels. We therefore create a new memory image in which the pixels which had more than 200 ADUs are affected by the 4095 value and those which had less than 10 ADUs are affected by the 0 value. This is used in the STRETCH function of the LYNXX camera's acquisition software.

- Second method. The introduction of low and high thresholds (sometimes called 'cuts'). Without touching the memory image values, the transcoding table is described as follows: we assign black to pixels whose value is between 0 and 10 ADUs, white for those between 200 and 4095, and a shade of gray for those between 10 and 200. In this representation, the low threshold is equal to 10 and the high threshold is at 200 ADUs.

Both methods lead to the same image on the screen, with far better contrast than the preceding one (figure 4.4):

FIGURE 4.3 A screen representation of the memory image of figure 4.2. The transcoding table used in this example is: black for zero and white for 4095 ADUs. Notice the weak contrast of such a representation.

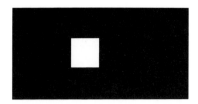

FIGURE 4.4 A screen representation of the memory image of figure 4.2. The transcoding table used in this example is: black for 10 and white for 200. Notice the remarkable increase in contrast compared to figure 4.3.

It is preferable to use the high and low threshold method than the STRETCH function since the latter modifies the image's contents in memory and, as we will later see, loses the image's photometry (we can no longer return to the original image in memory).

4.1.4 *The image on disk*

In order to save a CCD image after the computer is switched off, it must be stored physically. In amateur astronomy, most often, diskettes or the computer's hard disk are used. One can record either the memory image or the image on screen. In the first case, it is called the image on disk, and in the second case, the screen copy on disk. In both cases, a file named by the user is generated. On PC-compatible computers, the name must not exceed 8 letters, without spaces, followed by three characters called the extension, which distinguishes whether the file contains an image on disk or a screen copy. For example, the file name N7331-1. FIT could signify that it is the first raw image of the galaxy NGC 7331 and was registered in the FITS format.

The image on disk is therefore the recoding of the image in memory. The image is stored in the form of a file that contains all of the memory image's pixels. More precisely, it is the value, in ADUs, of each pixel that is recorded onto the disk. The file occupies as large a space on the disk as the memory image occupied in the computer.

Digital image software must be able to generate image files on disk that are decipherable by other software. The international astronomical standard file format for both amateurs and professionals is the FITS format (Flexible Image Transport System).

The FITS format allows the saving of, in addition to the image, image data in its heading (the size of the image, scale, orientation, comments, and so on). If one wishes to recover images using windows based software, we suggest saving the images in TIFF format.

Some software can save images in a compressed form. The compacting algorithm should, therefore, be checked to ensure that the pixels' exact values are kept. This is called 'lossless compression.'

4.1.5 *The screen copy*

The screen copy on disk consists in copying the computer screen's pixel values into a file. The pixels' color is first converted into numbers before being recorded onto the disk. The screen's dynamic range, being less than that of the pixel, means that only the image in memory contains all of the information. The screen copy must only be used to quickly present a visual result, but it does not enable one to keep all of the memory image's quantitative information. There are several screen copy standards on disk for compatible computers: they are the PCX, BMP, TIFF, GIF, EPS, TGA, and JPEG formats.

We can also copy the screen image using a printer. Ink jet printers are less expensive and properly reproduce the images if they have sufficient contrast. Unfortunately, most sky images are low contrast and ink jet printers can give a rather poor result on paper. Some modern ink jets are now much better and offer good contrast and high resolution. Dye sublimation printers allow nice reproductions. Unfortunately, such printers and the special paper which must be used are more expensive.

A final solution consists in taking a photograph of the screen. We have seen that the screen's dynamic range is lower than that of the image in memory, but a photographic film's dynamic range is, generally, even less than that of a screen! It is always very difficult to obtain a good photograph of a screen while at the same time showing the maximum detail of the image in memory. Our experience shows that it is necessary to lower the screen's contrast and display the images with a gray background before photographing the screen. Screens are not, in general, flat, and to limit the effects of distortion, we suggest placing the camera as perpendicular as possible to the center of the screen. Now, the further one is from the screen, the easier it is to produce an image. Therefore, a camera should be mounted on a tripod about 2.4 meters from the screen. At this distance, the camera must be equipped with a 200 mm telephoto lens.

The electronic beam which forms the image sweeps across the screen about 70 times per second. Each sweep generates what we call a frame. To produce a proper image, several frames must be integrated, which requires a time constraint on the photograph's minimum exposure time. Counting a minimum of ten frames, an exposure must therefore be taken at at least an 1/8 of a second.

Taking the above into account, one should use a film sensitivity of 100 ISO, with the lens set at $F/D=5.6$ and exposure times in the order of 1/2 to 1 second.

If one wishes to use a black and white negative film, exposed in the preceding conditions, a very soft paper must be used (grade 0 or 1).

4.2 Display functions

Generally, display consists in presenting on the screen a collection of useful information contained in the memory image. This information can be the image itself, graphs, or even a series of characteristic numbers. The displaying of this information is generated without modifying the memory image's pixel values. We must always remember that a badly displayed image does not necessarily mean that the image in memory is of bad quality.

The modification of the memory image's pixel values is done through image processing operations. We must not, therefore, confuse displaying and image processing: software capable of displaying an image in false colors or in a 3D perspective does not necessarily mean it is capable of performing bulk image processing!

Finally, CCD image acquisition software does not need to resort to sophisticated display options, classic display and the image histogram are enough to work efficiently. But for the moment, we will describe the different types of display.

4.2.1 *Classic display*

The most natural representation of an image is to display it on a screen in the form of a gray scaled pixel array. Each screen image's pixel corresponds to a pixel in the memory image (see section 4.1.3). We have seen above that it is necessary to define low and high thresholds between which the software constructs a transcoding table that will relate the screen brightness of a pixel to its ADU values.

A debate which has existed since the first display techniques were developed centers on knowing how many different levels of gray are needed between the low threshold and the high threshold to 'correctly' display an image. In fact, this depends greatly on the nature of the content of the image to be displayed. In general, only a 256 level grading allows an excellent quality restoration for all types of image. Beyond that, the human eye cannot differentiate between two adjacent levels.

Up until now, we have considered that the degree to which a pixel is illuminated is proportional to the value of the ADUs: this implies linear transcoder tables. There also exist 'logarithmic' transcoder tables, where the brightness increases rapidly for the lower levels and more slowly for the higher levels. The

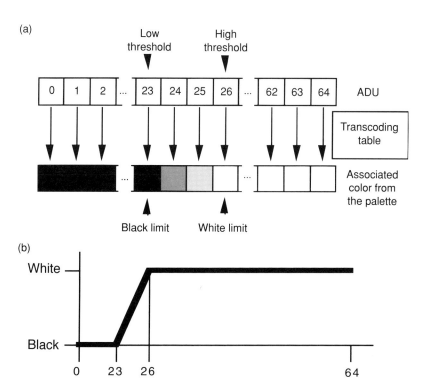

FIGURE 4.5 (a) The transcoding table, called the LUT, establishes the correspondence between the memory image's ADUs and the colors of the screen image. In the present case, the low threshold is equal to 23 and the high threshold is equal to 26 ADUs. The case shown here shows a gray gradation; we say that we are using a palette of 'gray levels'. We could also, without changing the transcoding table (that is, without changing the progressive brightness changing rule), use another palette (of 'red levels', for example). (b) The transcoding table described in (a) is represented here by a graph that allows us to find the correspondence between the ADUs and brightness.

logarithmic transcoder table is very useful for displaying images of galaxies or comets in order to distinguish details near the sky background while keeping details near the nucleus (see figure 4.6).

If a (linear or logarithmic) palette's black and white roles are reversed, a 'negative' image to displayed. If several gradations are produced between the low and high thresholds, a saw-tooth LUT is obtained, which allows one to display 'pseudo-isophote' images (see figure 4.7).

Aside from the black and white palettes, we can also ask the software to display images with another palette. For example, the black could be replaced with blue, and white with red. Between the low threshold (blue) and the high

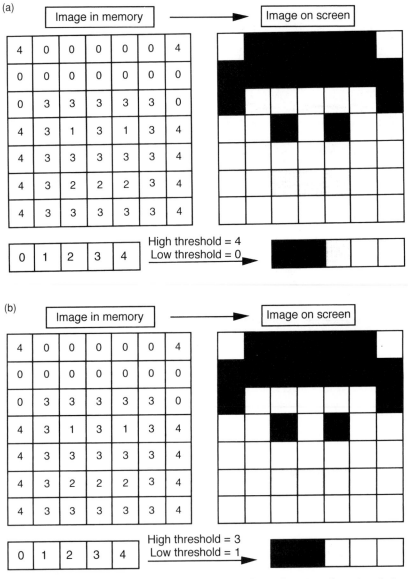

FIGURE 4.6 (a) On the left is a memory image digitized over five ADUs (from 0 to 4). On the right, the display is adapted to show useful details. (b) The same memory image as in (a). On the right, a higher contrast display is obtained by bringing the low and high thresholds closer together. In this case, notice how certain details are lost compared to the display in (a).

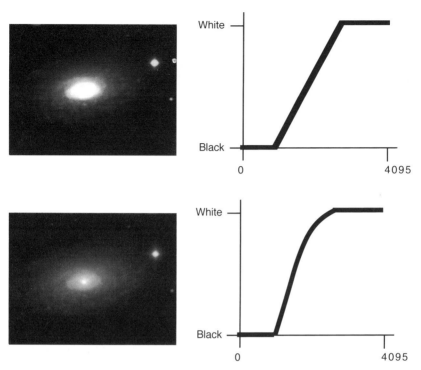

FIGURE 4.7 Top right: a linear LUT graph. bottom right: a logarithmic LUT graph. Top left: a memory image of the M63 galaxy visualized with a linear LUT. Bottom left: the same memory image displayed with a logarithmic LUT. Notice that the logarithmic LUT is very useful for making the internal galaxy detail appear while keeping the fainter extensions. CCD image: Bruno David, Alain Klotz and Gilles Sautot, Association T60.

(red) we could cover a range of colors reminiscent of a rainbow. This palette is often called a false color palette. We can create an infinite number of palette combinations in false colors.

We have seen that the screen image is commonly displayed according to the rule which states that one screen point represents one pixel of the memory image. Some software provides display functions that allow the modification of this ground rule. It becomes possible, therefore, to make one memory image pixel be represented by several pixels on the screen, or vice versa. Hence, it is possible to perform zooms or reductions. If we choose different enlargement factors on both axes, we can then 'normally' display a memory image from a camera with rectangular pixels. There are different enlargement algorithms, the best being, without a doubt, those of the B-splines, especially when we want to enlarge or reduce by a non-integer factor.

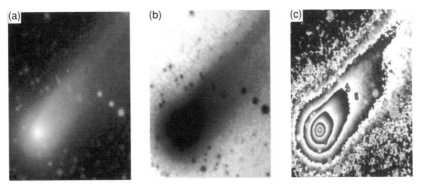

FIGURE 4.8 (a) The Swift–Tuttle comet, in 1992, displayed with a normal transcoding table. (b) The same image displayed with an inverse transcoding table. (c) The same image displayed with a saw-tooth transcoding. CCD image: Patrick Martinez, LYNXX camera and a flat-field camera of 500 mm, *F/D*=3.5.

To display an image, the user indicates the low threshold and high threshold values (in ADUs) between which the software will calculate the correspondence to give between a pixel's ADUs and the color to display it on the screen. This relation is called transcoding (or LUT for look up table). Transcoding tables are to be found in which progressions are, for example, linear, logarithmic or saw-tooth. It is also convenient to choose the type of palette used by the transcoding table; for example, gray graded, rainbow, or false color.

4.2.2 *Blink display*

One of the advantages of manipulating digital images is being able to display an image that has just been acquired, immediately after the exposure, and compare it to a reference image previously taken in the hope of discovering the appearance of a new object (a nova, asteroid, comet, supernova, etc.). The simplest comparison consists in displaying both images, one beside the other, and proceeding with a visual examination. This method is not infallible since it can be visually tiring when the image is large and there are a lot of stars.

Blink display consists in alternately displaying in a cyclic fashion two images in the same place on the screen. If one of the images contains one star more or one star fewer the alternating displays will induce a 'blinking' effect, which will allow the operator to immediately detect the presence of the new object.

It is necessary to recenter the images through an image processing operation (see section 5.4.1) before displaying them in blink mode. To display two images in blink mode, the software must be adjusted for the high and low thresholds so that the images are closely matched for brightness. In general, the alternating frequency of the images is adjusted interactively, by keyboard or mouse, during display.

4.2.3 *True color display*

The display of an image in true colors requires three images to be taken through three different colored filters: red, green, and blue (RGB). The simultaneous display of the three images on the screen, each one displayed with its own respective palette, is called a three-color process. Of course, each image must be recentered with respect to the others before they are displayed. We will discuss centering techniques in section 5.4.1 on image processing.

To display a tri-color image, three images must be in memory (RGB) and three low thresholds and high thresholds must be introduced.

We will see, in the section on photometry, how to assign weights to each of the three components. Anyhow, astronomical images are often very low contrast and increasing the color saturation is necessary. We can achieve this by tightening the display thresholds. We will see, later on, that a more radical technique consists in treating the three images by an HSI (hue, saturation, intensity; see section 5.3.3) system transformation before the three-color process display.

In general, software displays tri-color images with graphics modes allowing a display of no more than 256 different colors from a palette of 200 000 or 16 million colors. Before displaying the tri-color image, the software must choose the most representative colors for the image. This implies a simultaneous analysis of the three images to find the closest matching color. Several algorithms exist, relying, in general, on the histogram's density analysis (see section 3.1) of the three images. Luckily, since 1994, it is no longer rare to find graphics cards that allow the simultaneous displaying of 16 million colors on the screen. In the latter case, there is no longer any need to interpolate the colors from a limited palette and the tri-color image is better.

4.2.4 *Isophote display*

If we make an analogy between the memory image and a geographical map, the isophotes are to the ADU's intensity what contour lines are to altitudes.

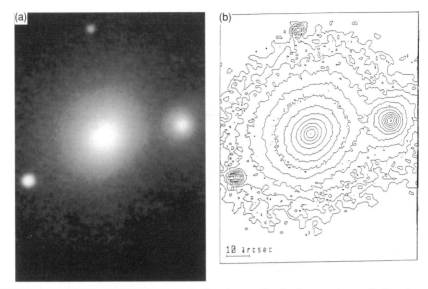

FIGURE 4.9 (a) An image of ARP 167 displayed normally. (b) the same image displayed by isophotes spread by half magnitudes. CCD image: Jean-Pierre Dambrine, Philippe Prugniel and Claire Scmitt-Darchy. 1 meter telescope at the Pic du Midi Observatory.

Therefore, we can display a screen image representing isophotes. We must also choose the number of ADUs between two isophotes. Although the principle is very simple, the isophote's tracing algorithms are very complicated, and there is bad software around that does not properly restore the isophotes. This can also serve as a good test to quickly judge the quality of the software.

The isophote representation can be superimposed very effectively onto the displayed image in normal mode. Indeed, the isophotes work best with circular symmetrical objects, such as comets or elliptical galaxies. Also, for these objects, it is convenient to use a logarithmic isophote distribution. Hence, we can reveal invisible structures in the normal image (jets within comets or isophote twists in elliptical galaxies).

4.2.5 Polarization card display

To measure the polarization of a celestial object, three images are taken filtered by a polarizer, oriented in a convenient manner. Then, as we will see in section 5.3.4, one extracts two images from the initial three images. These two images are the P image of the degree of polarization and the A image of the electric vector angle. The display of the polarization map consists in

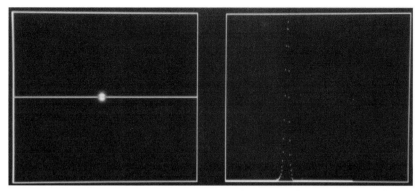

FIGURE 4.10 The left-hand image is a normally displayed star. On the right, the same image is displayed by the line cross-section which passes through the star's brightest pixel. The number of pixels that have a greater intensity than half of the maximum intensity of the star determines the image's seeing (eight pixels in this case).

displaying on the screen, for many image points, a line that represents by its length the degree of polarization and by its orientation the angle of polarization.

Generally, we choose to trace a line by a group of 9 to 16 pixels. The result on the screen is spectacular since we can directly see the appearance of magnetic field vectors in the studied object.

4.2.6 *Cross-sectional display*

All of the representations discussed up until now display the memory image as 'seen from above'. It is possible to represent a line or column of the image in memory seen from the side. That is, the displayed image represents, on the abscissa axis, the line or column pixels, and on the ordinate axis, the number of ADUs corresponding to each pixel.

The cross-section is very useful for determining a star's full width at half maximum. We display the cross-section on the line that passes through the most luminous pixel of the star. We count the number of pixels that have a higher intensity than half the intensity of the star's maximum. This number of pixels corresponds to the full width at half maximum, or FWHM. This is the number of pixels we can convert into arcseconds if we know the scale of the image (number of arcseconds per pixel). The FWHM corresponds to the image quality criterion which we call seeing, used by professionals. With a 20 cm amateur telescope, we can generally attain 5 second seeing during long exposures (or better if the site permits).

 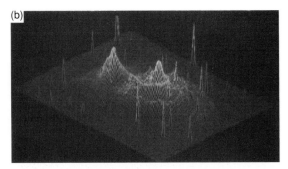

FIGURE 4.11 (a) A classic display of the galaxy M51. (b) The same memory image is displayed in three dimensions.

4.2.7 *Perspective display*

Perspective display (often referred to as 3-D, for three-dimensional) allows the memory image to be represented by a plane at any arbitrary angle through the same procedure as the cross-section, that is, the vertical axis represents the pixels' intensity. This time, the cross-section is applied to all the lines and all the columns. The result is often spectacular since we get the impression of vertical relief. Let us not be mistaken, though, for perspective display is more of a gimmick than a scientific tool! There are a multitude of possible projections, isometric or trimetric. The most important of a perspective display function's sophistications is to keep track of hidden lines.

4.3 Analysis functions

Analysis functions consist in extracting synthesized information from the memory image. The simplest analysis function is the cursor. After displaying the memory image onto the screen, the user moves a cross-hair over the image (with a mouse, for example). At each cursor position of the image on the screen, the software displays the value corresponding to the number of ADUs of the associated memory image pixel.

4.3.1 *Display help*

The most elementary analysis functions are useful to help the operator to quickly determine the low and high threshold values in order to execute a good display of the memory image. We have seen the cursor case, but there are others:

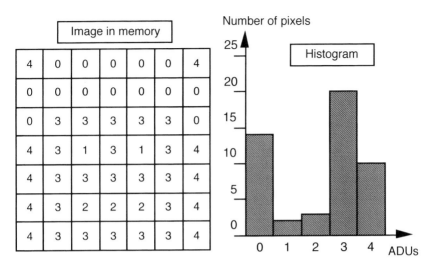

FIGURE 4.12 On the left, a matrix representation of a memory image. On the right, the histogram corresponding to the image on the left.

- We can ask the software to search for the weakest and strongest ADU values in the memory image. In planetary imaging, for example, we would have a direct idea of the high and low threshold values: the low threshold would be equal to the minimum value and the high threshold would be the maximum value. The image's maximum value also serves to check, at the end of the exposure, that no pixel was saturated.

- An image's histogram is a graph which represents in abscissa the ADUs, and in ordinate the number of pixels in the memory image that have the corresponding ADU. Figure 4.12 shows an example of a histogram. The graphical analysis of the histogram enables more precise information to be found in the case of ADU maximums and minimums.

The histogram is used in the LYNXX camera's software to indicate the minimum and maximum values of the STRETCH function described in section 4.1.3.

The method for determining the sky background is a quick but very useful one for rapidly determining a deep-sky image's thresholds. The method relies on the calculation of the average number of ADUs of the sky background. There are different ways of doing this, for example, by averaging the value of ten pixels situated near the sides of the image, for in a deep-sky image, the observed object is generally in the center of the field while 95% of the pixels to the sides of the image are on the sky background. To display a deep-sky image, digitized on 12 bits, it is usual to choose a low threshold for a sky background value of less than 20 ADUs, and a high threshold value for a sky background value of

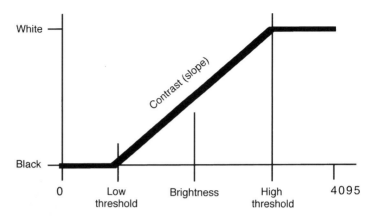

FIGURE 4.13 Luminosity is defined by the middle segment delimited by the low and high thresholds. The contrast is defined by the LUT.

more than 80 ADUs. Beware: this 'recipe' is not applicable if the image has undergone logarithmic scale processing.

The notions of high and low thresholds are not always properly linked to the eye's physiological sensitivity. To do this, some display software replaces the idea of thresholds by luminosity and contrast. In this case, if one examines the transcoding table's graphical representation, one can define the luminosity as the high and low threshold average and the contrast as the right slope that links the low and high thresholds. To make the problem easier, the software expresses the luminosity and contrast in percentages. If you do not understand the physiological aspect of these two values very well, turn the brightness and contrast buttons of your television set.

The cursor allows us to read the memory image's pixel value from its screen visualization. The histogram allows a graphical representation of the intensity frequencies in the image.

4.3.2 *Photometric analysis*

It is often necessary to determine the total intensity of a star in an image in order to further calculate its magnitude. For this, software is equipped with a moveable window in the shape of a rectangle or circle, whose dimensions are chosen by the user. The software displays, on the screen, the sum value of all the pixels located within this zone. The example of figure 4.13 illustrates the case where the sky background is equal to zero ADUs. In reality, there is always

0	0	0	0	0
0	3	6	3	0
0	6	18	7	0
0	3	8	4	0
0	0	0	0	0

FIGURE 4.14 The darkly outlined square delineates a zone of nine pixels, in which an extended star is found (in practice, we always choose the most extended zone so as to be certain to integrate the entire star). The software will calculate the sum of all the pixels within this zone. This integration operaton provides us with a value of 58. We will later see how we can deduce the star's magnitude.

a luminous contribution due to sky emissions or because of light pollution. We now need to make a measurement of the star, so we move the window from the sky background, center it on the star and execute a second measurement. We isolate the star's contribution by subtracting the two measurements.

The integrated intensity determination, on an extended zone, allows us to know the total number of photons that fell within this zone. It, therefore, becomes possible to determine a star's magnitude which is spread out over several pixels.

4.3.3 *Photocenter calculation*

In order to merge images from a same star field, we must, firstly, recenter them one according to another. To do this, we determine the precise location of a common star to each image. For this, software is equipped with a moveable window in the shape of a rectangle or a circle whose dimensions are chosen by the user. The software displays the barycentric coordinates of all the pixels inside the zone.

In the case of figure 4.15, the center of gravity position on the x axis is calculated in the following way:

$$\Sigma x = 3(3+6+3)+4(6+18+8)+5(3+7+4)=234,$$
$$\Sigma ADU = (3+6+3+6+18+8+3+7+4)=58,$$

and hence,

$$x = \Sigma x / \Sigma ADU = 234/58 = 4.03.$$

Likewise, we can calculate the position $y = 3.95$.

5	0	0	3	6	3
4	0	0	6	18	7
3	0	0	3	8	4
2	0	0	0	0	0
1	0	0	0	0	0
	1	2	3	4	5

FIGURE 4.15 Calculation of the barycenter (within the zone outlined in the dark square) allows us to determine the precise position of the star. On the matrix's periphery, the pixels' coordinates are indicated.

Figure 4.15 shows the barycentric position calculation of a star whose maximum pixel is the (4;4) position. In practice, we would choose a larger moveable window than the star to ensure finding the entire star within it. The calculation indicates that, in reality, the star must be considered centered at the accurate (4.03;3.95) position. It becomes possible, therefore, to do accurate astrometry with such a tool. Meanwhile, our experience shows that measurement dispersion by this method is about ± 0.03 pixel.

In practice, the user is asked for a value, above which the pixels will be taken into account in the barycenter calculation. This value is given the number of ADUs of the sky background plus 10 to 20 ADUs for a 12 bits digitized image.

The determination of a star's photocenter is useful for recentering images according to one another before adding them or before displaying them in tri-color.

---

5 Image processing functions

5.1 Image processing

Image processing consists in modifying the computer memory image's digital pixel values. Few processing functions are necessary to control the image acquisition. CCD camera control softwares are only generally equipped with a strict minimum of processing functions.

We can distinguish two major processing steps to apply to a raw image (which has just been acquired): the preprocessing and the processing itself.

We have seen, in chapter 3, that all raw images contain undesirable effects, inherent to the CCD system technology used. The preprocessing consists in 'cleaning' the raw image of these effects. Image processing is only efficiently applied to pretreated images.

5.2 Preprocessing

In this chapter, we define a raw image as any image output from the camera which has not undergone any automatic correction as some acquisition software does. In section 3.4 we established that the pretreatment of a raw image with an intensity of I (x,y) came to the following equation:

$$i(x,y) = \frac{[l(x,y) - l_d(x,y)]}{[l_f(x,y) - l_b(x,y)]} \times \text{constant}.$$

$i(x,y)$ is the pretreated image. $l(x,y)$ is the raw image. $l_b(x,y)$ is the precharge image, also called the bias or offset – it is the shortest possible exposure in total darkness. $l_d(x,y)$ is the camera's thermal response image, also called the dark. It is an exposure produced in total darkness, with the same time t and at the same temperature T as the preprocessed image; it also contains the precharge. $l_f(x,y)$ is the flat field image. On a flat field image, $i(x,y)$ is constant for all of the array's pixels. Constant is a normalization factor which is used to compensate the dynamic range loss which occurs during the flat field division. To each flat field, we associate such a factor. In general, it is equal to the average value of all the flat field's pixels.

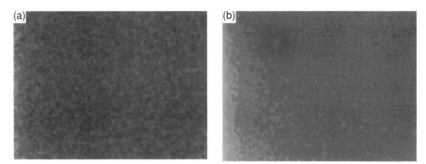

FIGURE 5.1 (a) A bias image shows noise. (b) The median bias image, obtained from 10 individual images, shows much less noise.

The pretreatment takes place in three very distinct steps: The dark subtraction (and the bias at the same time since it is included within the dark). Then, the flat field division (from which we have previously subtracted the bias image). Finally, the last step consists of locally changing the image in terms of the pixels or defective columns. This is the cosmetic correction step.

Over the course of each pretreatment operation, we improve the quality of the signal. But we also add noise, because the calibration images (bias, dark and flat field) are themselves noisy. Remember, noise is a random fluctuation of a signal around its average value. Before beginning the night's images pretreatment, it is best to synthesize calibration images, as noiselessly as possible. The following paragraphs, therefore, will be geared toward noise minimization techniques.

5.2.1 *The bias image*

The bias image is an image that intervenes, in terms of the flat field corrections, but also when certain special techniques are used during the dark subtraction. The camera's precharge image stays, generally, invariable regardless of the exposure time or the CCD detector's temperature. It is suggested that a bias image be made by applying a point to point median operation on a lot of ten individual precharge images (see section 5.3.2). Hence, we obtain a bias image with much less noise. Since the bias image evolves very little during the camera's life, we can, in general, do it once and for all.

If all of the bias image's pixels values are constant, excluding the noise, then a quick treatment consists of subtracting the constant value from the raw images rather than subtracting the noisy bias image.

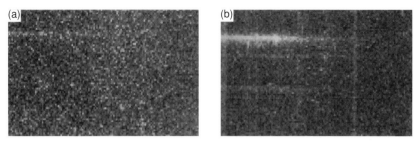

FIGURE 5.2 (a) A raw image of the spiral galaxy M77, exposed for 30 minutes at the T60 of Pic du Midi. (b) After the dark correction, the image clearly shows more details. Image: A.C. Afanassieff, D. Bardin, Y. Guimezanes, A. Klotz. Association T60.

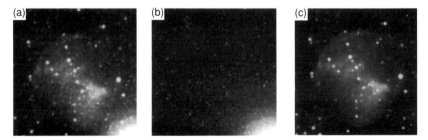

FIGURE 5.3 (a) A raw image of Messier 27 taken with a 5 minute exposure with an ST4 CCD camera focused on a 180 mm diameter telescope at $F/D=6$. (b) is the dark image with the same exposure taken in total darkness. (c) Their subtraction allows the suppression of most of the raw image's faults. Image: Jean Balcaen, Christophe Boussin and Vincent Letillois, Club Astronomique SCAA de Reims.

5.2.2 *Dark subtraction*

One of the major problems facing the pretreatment of raw images comes from the fact that the dark level depends on the exposure time and the camera's detector temperature. In the case of cameras cooled to -100 °C, the thermal charges are so weak that we prefer to remove a constant value from the image instead of subtracting the image from the dark in order not to add noise. This is, obviously, a rare case which is only encountered in some professional observations.

Flat fields, planetary images, solar images or double stars, often require a very short exposure time (less than 5 seconds) and it is often worth avoiding the dark removal, which would be very weak and add noise.

The classic method The most logical dark correction technique is the following: during the night, immediately after each raw image, we carry out a dark with the same exposure time as the raw image in order to keep the temperature conditions as close as possible to the raw image. This method has the huge drawback of spending half the observing time exposing on darks rather than observing the sky. Meanwhile, in practice, we can often be satisfied by using a single dark for several consecutive exposures with the same integration time. This dark could also be produced during the finding of the next object, which limits time loss.

A wide range of remedies can be taken to avoid producing a dark after each exposure. We will describe two of them.

The generic dark method This method consists of using a camera equipped with an integrated temperature probe on the CCD detector and functions in two steps. For the first step, if the camera is not temperature regulated, we carry out, once the camera has just been purchased, a series of generic darks with identical exposure times t_n but with different temperatures. Over the course of the night, note the acquisition temperature of each raw image and it will suffice to choose a dark exposed at the same temperature to pretreat the image. If the camera is temperature regulated, it suffices to carry out a single generic dark at a known exposure time t_n. The second step consists of creating a dark adapted to the exposure time t of the raw image by multiplying the generic dark image by a coefficient equal to the (t/t_n) ratio. This method reduces the noise by as much as the generic dark exposure time t_n is long.

Beware generic darks, as described above, must be corrected from the bias image (which is independent of the exposure time), before being multiplied by the (t/t_n) constant. Once the generic dark image has been multiplied by the proper constant, the bias image must be added to it before subtracting it from the raw image.

In order to avoid being obliged to produce a dark after each exposure of the sky, we can carry out a series of generic darks, which have been temperature and exposure time calibrated.

The dark optimization method In 1990, Christian Buil published an original method of pretreating raw images resulting from a single camera with a single generic dark, produced once and for all, with a long enough exposure time and an average operating temperature. This method is called, generally, 'dark optimization' and is integrated in Mips, PRISM and STAR softwares. It is the only method that allows a quasi-automatic raw image pretreatment. The foundation of this dark optimization method relies on the fact that a camera, whose thermal charge is insignificant during the array read, presents, regardless of exposure time or temperature, darks which are all pro-

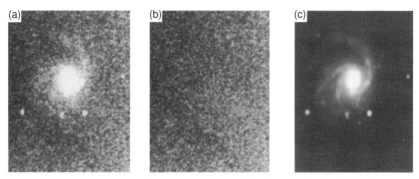

FIGURE 5.4 (a) A raw image of Messier 99, corrected for bias, in which a sky background area has been selected. (b) The optimized dark image shows, in the same areas, the same grainy aspect as in (a). (c) After the optimized dark subtraction, notice that the background is much more homogeneous than in (a). CCD image: Bruno David, Alain Klotz and Gilles Sautot, Association T60.

portional to one another. A single generic dark is necessary. It suffices to determine the multiplication constant to apply to it to create a dark to subtract from the raw image. We can assume that this method is applicable for cameras cooled below $-20\,°C$ for classic detectors and below $+5\,°C$ for MPP detectors.

In the first place, we locate, on the raw image, a zone in which there is only sky background, (which is almost always possible on astronomical images). This zone contains a signal due to the sky's brightness and another from the thermal charges. In general, for a camera cooled to $-30\,°C$, the noise texture of such a zone is mainly due to thermal load variations between the pixels since the photon and precharge noises are negligible and also because the sky's brightness is uniform enough on the scale of a few pixels.

The software searches for the multiplicative constant value that must be applied to the generic dark to create a dark that will best suppress, during the subtraction, the noise in the chosen zone of the image to pretreat. The determination of the optimization constant relies on a noise minimization method in the previously described image zone. In practice, the generic dark is the sum of at least ten darks of several minutes in order to minimize the noise. Be careful, before this sum, we subtract the bias from each dark. Before launching the optimization routine, it would be best to also subtract the bias image from each raw image. The optimization routine requests the window measurement and, eventually, the low and high limits of the multiplication constant to search for. At the end of the routine, the software displays the constant's value. Then, we multiply the generic dark by this constant and add the bias image to it. Finally, we subtract the optimized dark from the raw image. An application example is given in section 5.2.5.

The dark optimization method allows the correction of raw images from a single generic dark.

The nonlinearity of darks The mass arrival of MPP technology components has shown that some have a thermal signal that does not proportionally increase in relation to time (as in the case of the KAF-0400 CCD). This non-linearity greatly limits the use of the methods described in section 5.2.2. If we wish to use the generic dark or dark optimization methods, it is suggested that a dark produced in the closest conditions possible to the images to be pretreated be used.

The French amateur, Jean-Claude Pelle, has shown that we can carry out a complete calibration of a CCD camera pixels' thermal response by keeping track of thermal signal's non-linear variation in relation to time. For example, we can develop the $l_d(x,y)$ signal following this mathematical expression:

$$l_d(x,y,t) = l_b(x,y) + a_1(x,y)t + a_2(x,y)t^2.$$

To determine a_1 and a_2, we must first carry out a series darks at a t_1 exposure time, then synthesize the $l_d(x,y,t_1)$ image by doing the dark series's median. $l_d(x,y,t_2)$ is done the same way with $t_2 = 5t_1$. Since $l_b(x,y)$ is known (section 5.2.1), we can, therefore, determine $a_1(x,y)$ and $a_2(x,y)$ by resolving the system of two equations and two unknowns.

In this manner, we can use the dark optimization method described above for images from MPP CCD cameras. In this case, the method consists of finding the value of t which allows the minimization of the residual noise of the difference between the raw image and the optimized dark.

This method is not currently integrated in commercial image treatment software, but the ever increasing popularity of MPP technology CCD cameras will increase the demand for it.

5.2.3 *The flat field division*

Flat fields are common to all sky images taken with the same filter. Also, never forget that flat fields must be produced with the same focus and camera position as the sky images. If we do not move the camera's orientation on the telescope, we can keep the same flat field over several consecutive nights. However, it is always prudent to redo the flat field each night whenever possible. In order not to be annoyed by thermal charges, we suggest producing flat fields with exposure times between 1 and 5 seconds. In general, we produce about 10 flat fields at the beginning and end of the night.

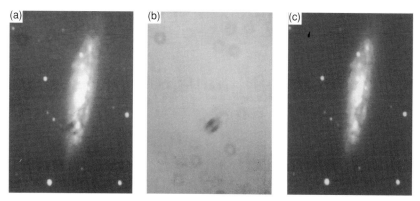

FIGURE 5.5 (a) A dark subtracted image, dark corrected, of Messier 108 (2 minute exposure). (b) The flat field image shows that there was dust on the CCD camera's aperture window. (c) The same image corrected for flat field showing a more uniform sky. CCD image: Bruno David, Alain Klotz, and Gilles Sautot, Association T60.

Prior to everything, we must begin by subtracting the bias image from all of the flat field images. Then, in order to diminish the noise, we create an average flat field by adding the 10 flat fields. An even more efficient method consists of, firstly multiplying each flat field by a constant in order to bring them all to the same average level and then synthesizing a much less noisy flat field by applying a point to point median operation on all 10 flat fields put on the same level (see section 5.3.2).

It is sometimes impossible to produce a flat field at the beginning or end of the night. We must then use more or less sophisticated methods to produce the flat fields from the night's images.

The simplest among them consists of targeting at least fifty pixels representing the sky background on one of the night's images, then asking the software to synthesize an average flat field by best adjusting a polynomial function on the targeted pixels. This method allows the correction of vignetting effects, but does not correct the presence of dust patches.

A second method consists of synthesizing the flat field by applying a point to point median operation on all of the night's images in which the sky backgrounds have been brought to the same level (by a point to point multiplication operation). This method improves with the number of images produced during the night. The major inconvenience of this method is the average level of these flat fields is weak and the pretreated image pixels value's uncetainty will be larger than with a twilight sky-produced flat field. In return, some cameras, most notably the 16 bits, can be affected by linearity errors and it becomes advantageous to correct images by flat fields which have the same average level as the images to be treated.

FIGURE 5.6 The image of Messier 64, dark and flat field corrected, still shows the presence of a defective line. Documentation: Alain Klotz.

To each flat field, we associate a numeric constant which is taken, in general, as the average value of all the flat field's pixels (from which we have previously subtracted the bias image). After having subtracted the dark from the raw images, we divide the image by this flat field and we multiply the result by the constant (the two division and multiplication operations must be integrated into a single function in order to be produced simultaneously).

Note that if we wish to produce tri-color imagery over the course of the night, we must use three flat fields (one for each filter).

5.2.4 *Cosmetic corrections*

Large CCD arrays often have defective pixels, lines or columns. Also, the Earth is constantly bombarded by cosmic rays and the CCD detector has the annoying property of registering their passing across the array: the result is the apparition of a few isolated, very bright pixels. Cosmetic corrections consist of removing these effects from the image.

Isolated pixel correction is a simple operation. It consists of placing the cursor on the defective pixel and asking the software to carry out the correction. The most commonly used algorithm replaces the defective pixel's value by the average of its four closest neighbors.

The correction of a defective column (or line) can be produced by replacing the value of each pixel of the column by the average values of the two pixels located in the adjacent columns.

In some cases, the pixel values contained in a defective column can be recovered if it is a 'cold' column. By definition, a cold column contains pixels

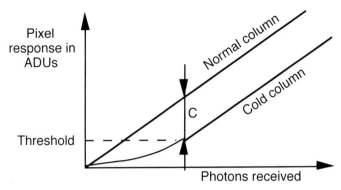

FIGURE 5.7 Representation of the response of a cold column pixel in relation to the response of a normal column pixel. The cold column correction needs one to know the threshold from which to add the C constant to the cold column's pixels.

that respond to a luminous signal with a delay. Beneath a certain light threshold, the pixels' response is weaker than the pixels of a normal column, and, above that threshold, the response return to normal. We see such a response described by figure 5.7. This phenomena is also called deferred charge. The correction of a cold column is possible if we know the threshold delay and the C constant to add to the cold column's pixels which have a higher average than the threshold. Below the threshold, we must interpolate the value to add.

5.2.5 *Preprocessing automation*

Software that has an integrated programming language allows the creation of pretreatment macro functions. In this paragraph, we will expose the reader to two macro functions that can be used, respectively, in the Mips and PRISM softwares. In this example, the entry images are all in the FITS format. The images would have been saved onto disk with the FIT extension. Also, the considered flat field image would have been previously corrected by the bias image.

The following programs must be written as is, in ASCII characters, with editing software. Once written, it would be convenient to save the program file, by giving it a name, onto disk: PRETRT.PGM, for example. The programs proceed to optimize the dark (OPT function for MiPS and the Blckopt for PRISM) and divide the flat field, but we will let the reader add the proper cosmetic corrections which are unique to each camera.

This program is written in the MiPS programming language and allows an automatic preprocessing sequence.

```
~Automatic Pretreatment Program
~for Mips software
~Images are in the FITS format

~Definition of file names
INPUT 'Name of the bias image:' %precharge STRING;
INPUT 'Name of the generic dark image:' %dark_generic STRING;
INPUT 'Name of the flat field image:' %ff STRING;
INPUT 'Name of the raw image:' %raw STRING;
INPUT 'Name of the pretreated image:' %pretreat STRING;

~Conversion of files from the FITS format to the MiPS format
%precharge_fits=%precharge&'.FIT';
IMPORT /in 1=%precharge_fits /out0=%precharge /file_type=fits;
%dark_fits=%dark_generic&'.FIT';
IMPORT /in 1=%dark_fits /out0=%dark_generic /file_type=fits;
%raw_fits=%raw&'.FIT';
IMPORT /in 1=%raw_fits /out0=%raw /file_type=fits;
%off_fits=%ff&'.FIT'
IMPORT /in 1=%ff_fits /out0=%ff /file_type=fits;

~Subtraction of the precharge from the raw image;
SUB /in0=%raw /in1=%precharge /off=0 /out0=%pretreat;

~Dark optimization
WINDOW /in0=%pretreat /out0=zone /x1=20/x2=40 /y1=20/y2=40;
STAT /in0=zone /flagdisplay=0;
%ave_i=v3;
%coefl=1.1*%ave_i/%ave_n;
%step=%coefl/%ave_i;
OPT /in1=%pretreat /in2=%dark_generic /x1=20 /x2=40 /y1=20 /y2=40;
/coefl=%coef1 /coef2=0 /stepcoef=%step;
%mult=%v1;
MULT /in0=%dark_generic /out0=dark /coef=%mult;
~Subtraction of the optimized dark;
SUB /in0=%pretreat /in1=dark /off=0 /out0=%pretreat;

~Flat field division;
STAT /in0=%ff /flagdisplay=0;
%ave=v3;
DIV /in0=%pretreat /in1=%ff /out0=%pretreat /norm=%ave;
```

```
~Displaying and saving of the pretreated FITS format image;
BG /flagdisplay=0 /in0=%pretreat;
%high=vl+50 ; %low=vl-20;
hl=%high ; ll=%low; %ey=100;
VISU /in0=%pretreat /hl=%high /ll=%low /ey=%ey /mode=1 /frame=2
/step frame=10;
%pretreat_fits=%preatreat&'.FIT';
EXPORT /in0=%pretreat /out1=%pretreat_fits /file_type=fits;
DELETE /in1=%precharge /confirm='Y';
DELETE /in1=%dark_generic /confirm='Y';
DELETE /in1=%ff /confirm='Y';
```

This program is written in the PRISM programming language and allows an automatic preprocessing sequence.

```
REM PRISM software automatic
REM pretreatment program
REM Images are in the FITS format

REM Definition of ifle names
Input 'Name of bias image: ';Pre$
Input 'Name of generic dark image: ';Dark$
Input 'Name of the flat field image: ';Ff$
Input 'Name of the raw image: ';Raw$
Input 'Name of the pretreated image: ';Prt$

REM Subtraction of the precharge from the raw image
Load Pre$
Tbuf 1
Load Raw$
Subt 0

REM Dark optimization
Tbuf 1
Load Dark$
Swap
Qview
Blckopt
Swap
ClearLine
QPrint 'Coef='
Qmult
```

```
REM Optimized dark subtraction
ClearLine
Swap
Subt 0

REm Flat field division
Tbuf 1
Load Ff$
stat
PutStat Max Ave Min
Print 'Max='; Max ' Ave='; Ave ' Min='; Min
Swap
Div Ave

REM Displaying and saving of pretreated image in the FITS format
Qview
Save Prt$
```

The MiPS software pretreatment program is initiated with the RUNPROG PRETRT command. The program requests the names of the associated files of different images. We enter only the names without the FIT extension. Then, the program runs itself. Finally, it displays a pretreated image and saves it onto disk.

In terms of the PRISM software pretreatment program, we should first select FITS file format by going through the menus: Option/Import/Format/Fits. The program is initiated through the menus: Option/Prog/Run and by choosing the PRETRT.PGM file. Note that we also have the choice of initiating the program by placing it in 'expert mode' and selecting the RUN PRETRT command. The program requests the file names associated with the different images. We enter only those names without the FIT extension. Then, the program displays the raw image. The visualization parameters are adjusted with the threshold bar at the right of the screen. Once the thresholds are adjusted, press the ESC key to continue. The program then displays a cursor on the screen in order to define a window onto the sky in which the dark optimization will take place. The window is defined by clicking on each of its two opposing corners. Finally, the pretreated image is displayed and saved onto disk.

We must, in the meantime, remain prudent while using an automated pre-treatment system. Some images do not lend to it well and give false results. The golden rule is to save the result under a different name than the raw image file and display it immediately after its pretreatment in order to diagnose the smallest problem quickly.

5.3 Point to point operations

By the title 'point to point' we refer to all transformations that create an image in which the coordinate pixel (x,y) will depend only on the same pixel coordinates (x,y) for one or several reference images.

5.3.1 *Arithmetical operations*

The arithmetic operation is, without a doubt, the simplest image treatment imaginable.

Here is a list of arithmetical operations that intervene with a single reference image (figure 5.8 illustrates some of them):

- Adding of a negative or positive constant (useful before carrying out a logarithmic operation on an image or after passing morphological filters)

- Multiplication of a whole or decimal constant (used during the flat field normalization before extracting a median image or for adapting a generic dark in relation to a raw image after the dark optimization step)

- Normalized logarithmic transformation (transforms the image's initial scale, proportionally to the number of photons received, into a proportional scale to the objects' magnitude. In practice, the software applies the logarithmic function to the pixels and multiplies their values by a constant, fixed by the operator, in order to return the intensities to a reasonable dynamic range)

- Transformation into an equalized histogram (useful for harmonizing the dynamic range of certain galaxy images, for example. In practice, this transformation modifies the intensities in order to obtain the most homogeneous histogram possible onto the entire image dynamic)

- Truncating of values above or below a certain threshold (useful for eliminating aberrant values that can take pixels after an image treatment operation)

- Binarization of an image (useful before carrying out a skeletization. Values below a certain threshold, fixed by the operator, are returned to zero, the others are put at 1)

Here is a list of arithmetical operations that involve at least two reference images:

- Addition of images (used during combination, for example)

- Subtraction of images (used during the dark subtraction over the course of the pretreatment)

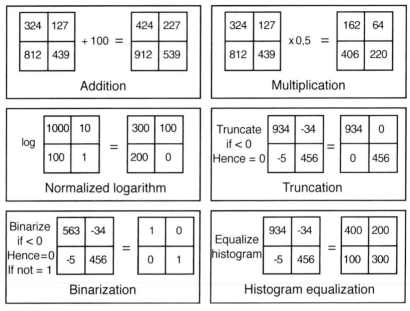

FIGURE 5.8 Illustration of the modification of a memory image's pixel values during various arithmetical operations which only involve the reference image.

- Image multiplication (rarely used except over the course of certain image restoration algorithms)
- Image division (used to correct flat field images during pretreatment)
- Any operation beginning from a reference image and a mask binary image (useful for applying the treatment only to certain image pixels)

Note that it is not unusual to find negative pixel values after an image treatment operation. Under the circumstances, we would choose to keep the negative pixels as they are or truncate them to attribute them a zero value or, perhaps, add a constant in order to return all of the image pixels to positive values.

All image treatment software is limited by maximum and minimum values which cannot be surpassed, otherwise, the treated image will be useless. This value is expressed in bit numbers, just as the CCD camera's analog-digital converter's saturation (refer to table 1.3).

A computer subtlety should be noted at this point: in terms of CCD cameras, a 16 bit digitalization signifies that the pixels' values are between 0 and 65535 ADUs since we are dealing with non-signed 16 bits. When it comes to image processing, the 16 bits format takes into account the pixel values between -32768 and $+32767$; we call this signed 16 bits. Hence, we may have a few

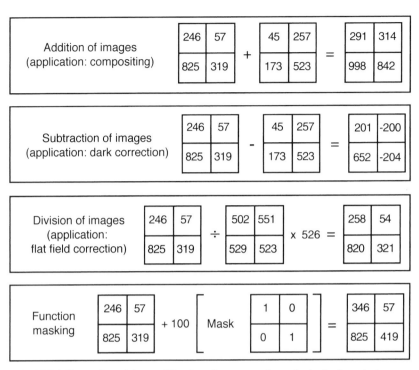

FIGURE 5.9 Illustration of the modification of a memory image's pixel values during various arithmetical operations which involve two reference images.

unwanted surprises during the 16 bit image read by software which does not take into account the non-signed 16 bit format. In the latter case, the pixels, which have a value above 32767 ADUs, could be systematically attributed an aberrant negative value.

5.3.2 *Statistical point to point operations*

We concluded in sections 5.2.1 and 5.2.3 that it is interesting to produce pre-charge images and synthetic flat fields by applying a median operation on a portion of different images. The median value of a series of numbers is obtained by classifying the numbers in increasing order, then choosing the median range classified number (=the middle one). For example, for the series of numbers (5; 7; 3; 8; 2), the increasing classing is (2; 3; 5; 7; 8). The median value, therefore is 5. This method is more efficient for eliminating aberrant values than a simple averaging.

The median's point to point operation could advantageously replace the sum

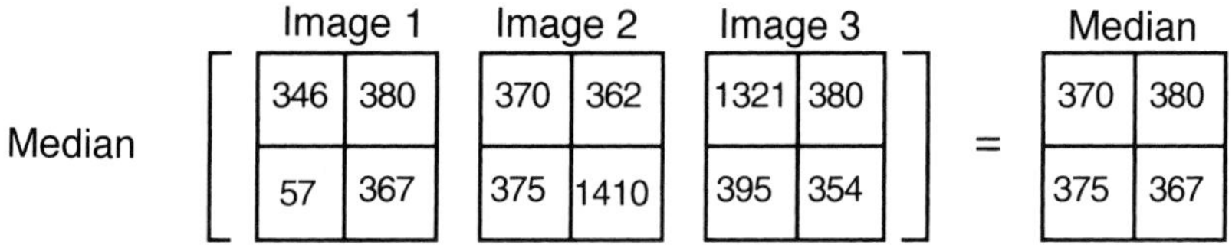

	Image 1		Image 2		Image 3			Median	
	346	380	370	362	1321	380		370	380
Median							=		
	57	367	375	1410	395	354		375	367

FIGURE 5.10 The median image is synthesized from three individual images. Each image contains a pixel with an aberrant value with respect to the average of around 370 ADUs. Notice that the median image allows the elimination of aberrant values; this is because it is used over the course of the preprocessing phase to create the flat field image.

operation during a combination when the signal is large. Indeed, if we wish to combine images with the same exposure time, we easily show that a median image's synthesis allows us to obtain a less noisy image than a classical summation of the combined image. Also, we can automatically eliminate most of the cosmetic faults.

The median operation's principle application is to synthesize a flat field image from about 10 images. Therefore, we efficiently eliminate defects on the individual images.

5.3.3 *Transformations in color space*

The acquisition of a CCD color image consists of taking three images of the same object, with red, green and blue filters. In these conditions, we say that a trichromatism is carried out in RGB color space. Unfortunately, celestial objects, such as galaxies or the moon often give rather deceiving results, even with the tightening of the visualization thresholds. Indeed, the color contrast is often very weak since the pixels' values appear 'too' correlated between the three RGB images.

The color geometric transformation is the most used of the point to point trichromatism image treatment operations. Starting from the three RGB images, we synthesize three new images defined in HSI space. The H signifies hue, S saturation and I is for intensity. There are several definitions for HSI space. We will address two of them here. The first is: the intensity image I is equal to the combination of the three RGB images. The number of ADUs of a pixel from an H hue image is as close to 0 as the color blue is above the color red. The saturation image S indicates the green saturation rate of each pixel. The number of ADUs of one pixel from the S image is as close to 0 as its color is green. Mathematically, we arrive at the following connection:

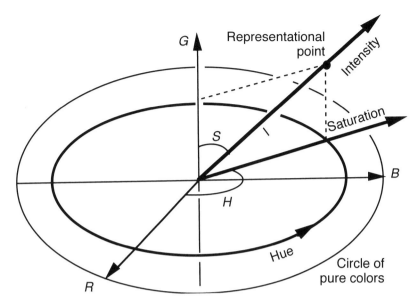

FIGURE 5.11 A graphical representation of the HSI coordinates with respect to the first RGB system described in the text.

$$H = \arctan\left(\frac{R}{B}\right),$$
$$S = \arccos\left(\frac{G}{\sqrt{B^2 + G^2 + R^2}}\right),$$
$$I = \sqrt{B^2 + G^2 + R^2}.$$

The second definition of HSI space also considers that the intensity image I is equal to the combination of the three RGB images. Contrarily, the S and H quantities flow from the Cartesian conversion coordinates R, G, B into polar coordinates by a rotation around the achromatic axis (R=G=B). Mathematically, we arrive at the following connections:

$$H = \arccos\left(\frac{2(R - 2G + B)}{\sqrt{B^2 + G^2 + R^2 - RG - RB - GB}}\right),$$
$$S = \arccos\left(\frac{B + G + R}{\sqrt{B^2 + G^2 + R^2}}\right),$$
$$I = \sqrt{B^2 + G^2 + R^2}.$$

After an HSI space transformation, it becomes impossible to easily modify the

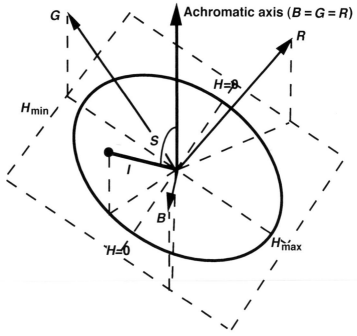

FIGURE 5.12 A graphical representation of the HSI coordinates with respect to the second RGB system described in the text.

color balance by activating the hue image or increasing the saturation. Then, we carry out an HSI space transformation toward RGB space before doing the trichromatism visualization.

We can imagine other types of geometric color transformations, for example, in cylindrical coordinates, but the HSI space is undoubtedly easier to interpret. Let us cite, for example, the principle component transformation method. It is a method that determines the most appropriate transformation by analyzing the RGB images' data. It is a complicated method although particularly efficient and can be extended to any number of colored composites.

5.3.4 *Transformations in polarization space*

To produce an object's polarization map, we place a polarizing filter in front of the CCD camera, located at the Cassegrain focus of a telescope. Indeed, on a Newton assembly, the light's reflection on the 45° mirror changes its polarization. During the acquisition, we produce three images of the object with the polarizing filter oriented along the relative position angles 0°, 60°, 120°. These images are noted, respectively, I_1, I_2, and I_3. The point to point transformation, allowing us to obtain the P image of the amount of polarization and the A image of the electric vector angle, are defined by the following relation:

$$k=2/3(I_1+I_2+I_3),$$

$$i=8/9\sqrt{(I_1-I_2)^2+(I_2-I_3)^2+(I_3-I_1)^2},$$

$$s=\sqrt{\frac{4}{3}}(I_2-I_3),$$

$$c=2I_1-k,$$

$$P=\frac{i}{k}$$

$$A=1/2\arctan\left(\frac{s}{c}\right).$$

There is another method that uses 4 starting images obtained with the filter rotated 45° between each component. In this case, starting from I_1, I_2, I_3, and I_4 we deduce P and A from the following relation:

$$P=\frac{2\sqrt{(I_1-I_3)^2+(I_2-I_4)^2}}{(I_1+I_2+I_3+I_4)},$$

$$A=1/2\arctan\left(\frac{I_2-I_4}{I_1-I_3}\right).$$

Caution! In this case, we can note that $I_1+I_3=I_2+I_4$. It suffices, therefore, to do the acquisition of the first three images and synthesize the fourth by carrying out point to point addition and subtraction operations on the first three. We have seen in section 4.2.5 how to display the polarization map on the screen defined by the P and A images.

5.4 Geometric transformations

The Geometric transformation possibilities of images is endless. The simplest is the flip-flop, which allows the reversal of the top and bottom or right or left. We can also simply extract a window from the image.

5.4.1 *Translations, rotations, and mosaics*

Translations A translation consists of 'sliding' all of an image's pixel values. In practice, image treatment softwares request the translation value according to the x and y axes. We are often constrained from producing fractionary pixel number translations (during compositing, for example).

The most current of translation applications is compositing, an indispensable operation for diminishing image noise as a result of poorly cooled cameras. Compositing consists of recentering and adding several pretreated images of the same object. The recentering is done through translations. The number of translation pixels is determined by the centroid method described in section 4.3.3.

The advantage of compositing is that it reduces noise by averaging and allows the quick locating of defective pixels by comparing individual images. Meanwhile, if we wish to detect the faintest magnitude possible, it would be better to composite 5 images of 10 minutes than 50 images of 1 minute. Indeed, if a 20th magnitude star generates 2 ADUs on 10 minute exposure images, the compositing of 5 exposures will show the star at 10 ADUs and can be detected if the image noise is not too strong. Contrarily, the same star at 20th magnitude, generates only 0.2 ADUs on images exposed for 1 minute. Since the camera digitizes on integers, the star will be attributed a zero value on the 50 images to composite and will, therefore, never be seen! In conclusion, the compositing of 50 images at one minute exposures will strongly diminish noise but will not produce an appreciable detection gain.

Rotations Rotation consists of 'turning' the image at a certain angle from its center. In practice, image treatment software requests the coordinate values of the rotation center and the angle. Note that it is possible to execute a rotation around a center which is outside the physical space of the image.

It is sometimes interesting to compare two images from two cameras fixed on two different telescopes. We are presented with two images whose scales and orientations are different. In this case, we must rotate one of the images to give it the same orientation as the other. Generally, rotation is not enough; it is therefore, necessary to carry out a translation and enlargement or reduction (see section 5.4.2). Practically, with performing softwares, matching the two images is done by, firstly, pointing to a series of details that are common to both images, generally stars, then applying a registration function that automatically carries out the proper rotation, translation and dilation on one of the two images.

Mosaics The CCD detector's field is sometimes too small to observe an entire extended object. We carry out, therefore, several exposures by moving the observed field between each image while keeping at least two or three common stars along the side of the fields. We synthesize a single large field by assembling all the small individual images like a mosaic. This operation consists of translating all of the images into registration with each other over the course of the assem-

FIGURE 5.13 (a) An image of NGC774 taken at the T60 of the Pic du Midi Observatory. (b) The same object taken with the focus of a T250 at the Pises Observatory. Matching the correct scale and orientation allows the effective compositing of the two images (c). CCD image: Cyril Cavadore and Jean-Marie Lopez, Association T60 and Société Astronomique de Montpellier.

bly. The translation value between two adjacent images is determined by the position measurement (x,y), determined by the centroid, of a common star in two images.

Before beginning the translation procedure, all of the individual images should be brought to the same dynamic intensity, the sky background must be within 1 ADU in order to avoid the visible discontinuities (we reach our goal through local subtractions and/or multiplication by a constant). Of course, the individual images must be perfectly corrected with the flat fields!

For merging overlapped areas the operator must be able to choose between several methods: uniquely keeping the pixels of one image or another, averaging the intensities of common pixels, keeping only the pixels having the maximum or minimum value, . . . The choice of one method over another is only guided by the final mosaic's aesthetics. With that said, there are few general rules.

Permutation We will see, when reviewing double star image processing, that we must have a cross-permutation of the quadrant images. The upper left quadrant must reverse itself with the lower right one and the upper right with the lower left. Hence, the four corners of the initial image will be found in the center of the permutated image and vice versa.

5.4.2 *Enlargements and reductions*

We can assemble pixels by groups to obtain a reduced image, this is digital binning. Contrarily, it can be interesting to zoom in order to enlarge an image

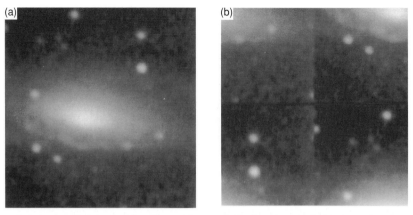

FIGURE 5.14 This image of the galaxy NGC7331 is displayed normally in (a). In (b) it has undergone a permutation transformation of the four quadrants, as described in the text. LYNXX CCD image: Juan Guarro (Barcelona).

FIGURE 5.15 (a) An image of the famous Clown nebula NGC2392. (b) The image is enlarged through a simple pixel enlargement. (c) The spline interpolation enlargement obtains a smooth result and does not generate aberrant details.

detail. In this case, the software will generate supplementary pixels through interpolation. The best algorithm, in this area, is the B-Spline interpolation. Indeed, this algorithm creates points by interpolations of 3rd degree polynomial functions with a constraint of null second derivative of the polynomial at the level of the original pixels. This has the interesting effect of smoothly enlarging the image without adding aberrant information.

In terms of binning, we have seen that we can produce analog binned images by combining the CCD camera's photosensors. In terms of the image treatment, we can carry out a digital binning by regrouping the memory image's pixels. For reasons mentioned in section 2.5.5, it is preferable to carry out binning at the image acquisition level, except with spectral images (except during wavelength centering) since it is preferable to firstly correct the contribution of the point to point faults (cosmetic and defective pixels) before

FIGURE 5.16 (a) A spectral image of the variable star V CVn at magnitude 8.6. The wavelength dispersion is horizontal and extends from 6481 Å to 6626 Å. Despite a 1 hour exposure, the signal to noise ratio is very low. (b) after a digital binning operation, this ratio is considerably increased. The Hα line becomes visible, as a very narrow line toward the spectrum's center. Images: Daniel Bardin, Valérie Desnoux and Michel Espoto; Association T60.

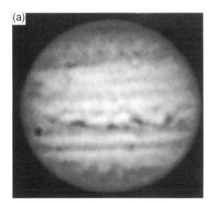 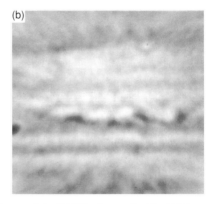

FIGURE 5.17 (a) An image of the planet Jupiter on 9 April 1993, observed with a T400, $F/D=19$. This image was previously treated with an unsharp mask (see section 5.5.1). The transformation to a cylindrical projection: Gino Farroni, Saint-Avertin (Indre et Loire).

doing the binning. In this case, we obviously lose a little signal to noise ratio but we are certain of being able to remove the noise interference contribution.

5.4.3 *Anamorphs*

An anamorph is a different geometric transformation according to directions. All of the following transformations belong to this category. For example, when we use a rectangular pixel CCD camera, we are looking to enlarge the image on a single axis in order to find the result as it would have been obtained with a square pixel camera.

Another type of anamorph consists in 'developing' the image of a planet on a plane surface, or 'planisphere'; it is a cartographic projection operation. There

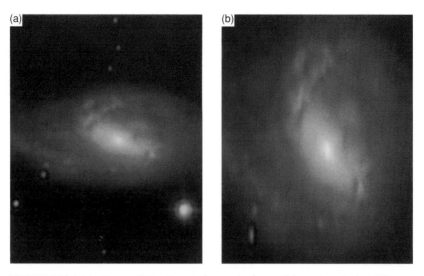

FIGURE 5.18 (a) An image of Messier 66 taken at the focus of the T60 at Pic du Midi.
(b) The vertical scale has been enlarged by a factor of 2.5 in order to show the morphology of the galaxy from a 'top view'. The artifacts around the stars are due to Lucy–Richardson algorithm restoration processing (see section 5.6.3). Image: B. David, A. Klotz and G. Sautot, Association T60.

are a multitude of possible projections, the simplest being cylindrical projection. In this case, the planisphere image is composed in the following manner: x axis represents the longitudes and the y axis represents the latitudes. If we choose such a projection with a scale of 1 degree per pixel, the planisphere will be contained within a 360×180 pixels image format. Let us also note that the American organization, United States Geophysical Survey (USGS) has established a standard which allows the minimum possible deformations in order to draw up planetary surface maps. We must also be able to go from the planisphere to a 'globe' representation, called oblique orthographic. This allows us to view the true image of a planet from its reconstituted planisphere.

Another interesting transformation consists of representing the projected image in polar coordinates. On the initial image, a center must be defined around which the polar development will take place. Generally, it is a galaxy's or comet's nucleus. After the transformation, the image is represented in the following way: the abscissa axis represents the θ position angle and the ordinate axis represents the distance ρ from the center (see figure 5.19). The benefit of such a transformation resides in the morphological analysis of the galaxies' curling arms or in the study of jets distribution in a comet and we will see, in section 6.2.2 how it influences the measurement of double stars.

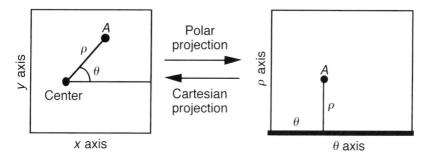

FIGURE 5.19 On the left is the initial image (in Cartesian projection) on which were placed polar reference coordinates and a point A with coordinates (ρ,θ). On the right, the polar projection transformation shows the way to project point A with coordinates (ρ,θ). The dark line, at the bottom of the polar projection, symbolizes the location of the center defined in the Cartesian projection.

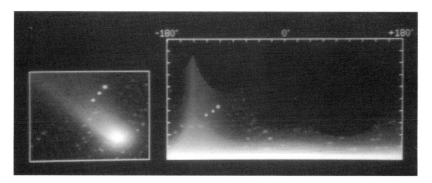

FIGURE 5.20 An image of the Swift–Tuttle comet taken with a LYNXX camera placed at the focus of a 500 mm focal length flat field objective. On the left is a Cartesian projection (normal representation). On the right, after a polar projection around the center, the cometary tail becomes vertical. CCD image: Patrick Martinez.

5.5 Convolutions

The spatial convolution of an image consists in regenerating, for each pixel, a value that takes into account the neighboring pixels' values. The frequency convolution of an image arrives at the same result but needs transformation into the Fourier domain. The notion of convolution is closely linked to filtering problems. All filters do not necessarily keep the initial image's photometry, therefore it is not linear. The reader must be aware of this.

10	10	10	10	10
10	10	10	10	10
10	10	200	10	10
10	10	10	10	10
10	10	10	10	10

FIGURE 5.21 An initial memory image of a star occupying a single pixel.

5.5.1 *Spatial convolutions*

Spatial convolutions are image treatment operations that consist of modifying the pixels' values in relation to their neighbor's values. We distinguish high-pass filters, intended to increase the local contrast, and low-pass filters, intended to spread out the images, that is to say, make them unsharp.

High-pass filters The high-pass filter is an image convolution which improves the local contrast. Let us describe its principle: figure 5.21 represents a memory image with a star occupying a single pixel, with an intensity of 190 ADUs, on a background of 10 ADUs.

In the original image, the pixel containing the star has an intensity of 200-10 ADUs while the sky background is at only 10 ADUs. We can define its contrast by the following calculation:

$$\text{contrast} = (200 - 10)/(10) = 19.$$

Let us now apply a high-pass filter: for a given pixel, we calculate the following number: 5 times the ADU value of the pixel minus the value of each of the four neighboring pixels (higher, lower, left and right). Therefore, we create a new image by applying this for all image pixels to convolve (see figure 5.22).

Obviously, we notice, that it was not possible to carry out a convolution on the border pixels since they have no neighboring pixels (where an X was placed) but this represents an insignificant loss on a large image. In the filtered image, the star represents a $(960 - 10)/10 = 95$ contrast, or a gain factor of 5 of contrast! We also notice the apparition of a few negative pixels around the star. They permit the conservation of the image's global photometry. In practice, we eliminate them by applying, to the filtered image, a point to point truncating

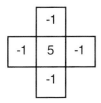

X	X	X	X	X
X	10	-180	10	X
X	-180	960	-180	X
X	10	-180	10	X
X	X	X	X	X

FIGURE 5.22 The image of figure 5.21 is convoluted by a high-pass filter. Notice that the star has better contrast. The crosses which appear on the image's periphery signify that we cannot calculate these pixel values. In practice, the images are sufficiently large that this loss goes unnoticed.

	-1	
-1	5	-1
	-1	

FIGURE 5.23 The convolution matrix of a high-pass filter.

FIGURE 5.24 (a) A raw image of the Moon exposed for 1 second at the $F/D=10$ focus of a Celestron 8. (b) The image convoluted by a high-pass filter, which considerably increases the contrast. CCD image: P. Martinez.

operation which resets all of the negative pixels to zero. Please note that we lose the photometric calibrations and the filtering is no longer linear.

The high-pass filter operation described above is also denoted by the graphic shown in figure 5.23, called a convolution matrix, which allows the quick visualization of the values and arrangements of the different coefficient multiplicators to apply.

The high-pass filter which has just been described is not unique. There is a complete range of filters that are different in their efficiency. Note that only the filters whose coefficient sum is equal to 1 are linear, that is, they keep the photometry. High-pass filters are particularly efficient on lunar images.

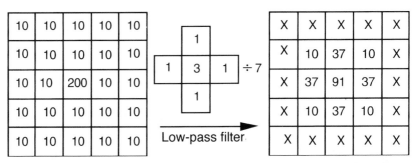

FIGURE 5.25 The use of a low-pass filter spreads the image, making it hazy. This technique can be used to create a mask for the unsharp masking technique.

Low-pass filters The low-pass filter family is used to render images unsharp. The result of a low-pass filter is a smoothing of the initial image. Figure 5.25 shows that the memory image star from figure 5.21, convolved by such a filter, spreads onto its neighbors. Since the convolution array's coefficient sum of figure 5.25 is equal to 7, we divide the result by 7 to keep the initial dynamic range and linearity. There are convolution arrays whose coefficients follow a Gaussian distribution; they are abundantly used during unsharp masking treatments.

It is sometimes useful to apply a low-pass filter on only the pixels near the sky background's value in order to diminish noise. It suffices to indicate, to the software, the value, in ADUs, beneath which the filtering will be applied. Hence, we can produce isophote visualizations in a much 'cleaner' manner than with the initial image. In this case, it is a non-linear filtering.

Low-pass filters have the habit of suppressing an image's small, high-contrast details. With that in mind, we can try to suppress defective pixels (cosmic rays, etc.) with the help of a low-pass filter. Indeed, it is preferable to use a median filter. This filter is not, strictly speaking, a convolution filter, but its functioning is very close: for pixel coordinates (x,y), we isolate the five values (in ADUs) associated with the pixel coordinates (x,y) $(x-1,y)$, $(x+1,y)$, $(x,y-1)$, $(x,y+1)$. We sort, in increasing order, the five numbers and extract the median value (=the third value sorted). We attribute the median value to the pixel coordinate (x,y). This method is very useful for automatically removing an image's aberrant points, while removing much less detail than a simple low-pass filter.

Finally, the 'out-range' filter, carries out the average of the 8 points surrounding the pixel (x,y). If the absolute value of the difference between the value of point (x,y) and the average is above a certain threshold, we replace, therefore, the value of pixel (x,y) by the average, otherwise we do not touch it. The effect is, generally, better with a simple median. Caution, out-range and median filters do not keep an image's photometry, they are non-linear!

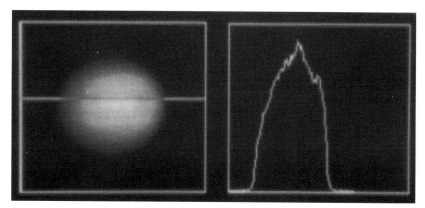

FIGURE 5.26 On the left, a raw image of Jupiter. On the right, the cross-section of the image shows that the planet is mainly composed of a component in the shape of a bell on which small details are revealed. As is, it is impossible to find a display that reveals these details with sufficient contrast. The unsharp masking technique consists in amplifying these details. CCD image: Gino Farroni.

Initial image	Unsharp mask	Detailed image

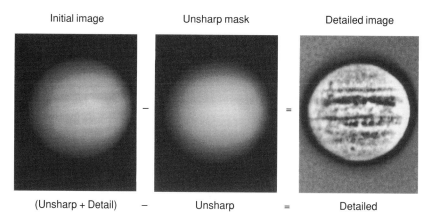

(Unsharp + Detail) —	Unsharp =	Detailed

FIGURE 5.27 The first step in unsharp masking consists in extracting the initial image's details. The detailed image is, unfortunately, too 'flat' to be presentable. CCD image: G. Farroni.

The unsharp mask technique The main convolution application for a low-pass filter is the unsharp masking technique, mainly used in planetary imaging. The principle consists of subtracting an image, convolved by a low-pass filter, from the original image. Indeed, we can assume that a planetary image has an unsharp component (the disk) on which we find details (bands, etc.). To extract details, it suffices to subtract the unsharp component from the initial image. This unsharp component is obtained by applying a very strong low-pass filter to the original image.

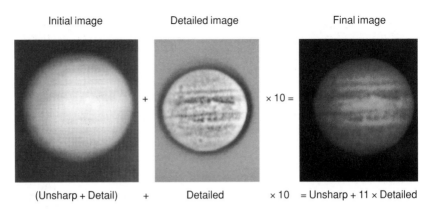

FIGURE 5.28 The second step of the unsharp masking technique consists in multiplying the detailed image by a coefficient between in general, 3 and 10, and then adding the whole to the initial image to give it a spherical aspect.

The unsharp masking technique allows the amplitude increase of an image's details. Therefore, we can considerably increase the contrast of planetary images.

Morphological filters We classify, in this family, filters which allow us to detect the contours of well-defined objects. Mathematically, this corresponds to calculating the first derivatives (gradient filters) and the secondary derivatives (Laplacien filters) of the initial image. Figures 5.29 and 5.30 show examples of the convolution matrices coefficient values associated to these filters. There are a large variety of contour filters. Among them, for example, are the Sobel filter and the Prewitt filter (revealing fine structures in galaxy arms). All of these filters' sums of their coefficients are equal to zero, they are, therefore, non-linear. In practice, after the application of a filter which has a null sum of its coefficients, we add a constant value to all of the image's pixels to return them to positive intensities. Skeletization is another method of analyzing the morphology of an object in an image. Figure 5.31, which shows the skeleton of an object, is defined by a network of lines formed by the center of inscribed circles tangent to, at least, two sides of the objects. It is rare to find programmed algorithms in astronomical image processing software capable of extracting skeletons via this method.

In general, the algorithms used attempt to dilate-erode an object. The filters associated with these operations are composed of a convolution matrix which transfers to the central pixel the minimum pixel value of the matrix to erode, followed by a dilation filter; it is called a 'closure'. When applied with the dila-

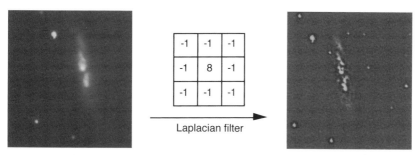

FIGURE 5.29 The Laplacian filter is used to reveal the contours of objects on an image. In this case, the active regions of the galaxy Messier 82 are revealed. CCD image: C. Boussin, J. Balcaen, V. Letillois. T180, $F/D=6$, ST4 camera, composite of two 5-minute exposures. Association Sciences et Cultures en Champagne Ardennes.

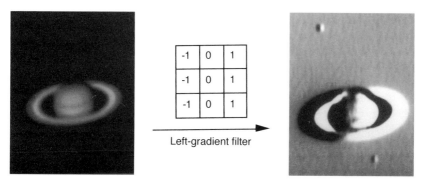

FIGURE 5.30 The gradient filter simulates a sharp directional shadowing. It suffices to change the sign of the matrices' coefficients to simulate lighting from the right. The gradient is useful for removing the masking continuous component and extracting faint details. In this case, we can reveal Saturn's satellites which were almost imperceptible on the initial image. CCD image: A. Klotz and J.P. Dambrine.

tion filter followed by an erosion filter, it is called an 'opening'. Obviously, these filters are not linear.

Skeletization algorithms result, generally, in binary images. In terms of their application, they can be used, for example, in tracing the spiral arms of galaxies.

5.5.2 *Frequency convolutions*

The previously described convolutions are produced in the 'spatial' domain since they act directly onto the image to be treated. Another method consists of transforming the initial image into two new images through a Fourier

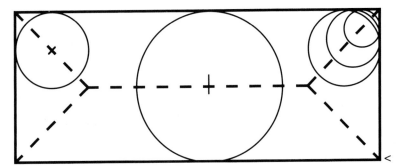

FIGURE 5.31 The rectangle in dark solid lines has, as a skeleton, the network of broken lines. The skeleton corresponds to the center of circles tangent to at least two sides of the rectangle.

transformation. It is a purely mathematical transformation which will not be described here because, in practice, it suffices to call it a transformation function and patiently await the results. In this book, we will commit ourselves, rather, to understanding the practical sense of this transformation.

Fourier transformations Starting from a spatial image, the Fourier transformation provides two images, which are called; the real frequency image and the imaginary frequency image, respectively denoted as R and I. The physical contents of these images only becomes clear after another transformation converts the R and I images into two other images called the amplitude spectral image and the phase spectral image, respectively noted as A and P. Figure 5.32 shows how these different image manipulations are organized. Be careful, in the frequency domain the transformation into polar coordinates has nothing to do with what has been defined in section 5.4.3!

The amplitude spectral image takes into account the intensity variations (in ADUs) on the spatial image. For example, if all spatial image's pixels had the same value, the amplitude spectral image would be composed of a single non-null pixel in its center. The central pixel corresponds to a null frequency variation on the spatial image, that is, the average value of all of the spatial image's pixels.

Let us add, to the continuous background of the spatial image, a noise across the entire image represented by random intensity variations between adjacent pixels. The amplitude spectral image will always have its central pixel non-null, but we will also notice that adjacent pixels on all four sides of the amplitude spectral image will also have non-null values. The amplitude spectral image's peripheral pixel zone, therefore, characterizes the presence of very rapid intensity variations on the spatial image. These rapid variations are called the high-frequency components.

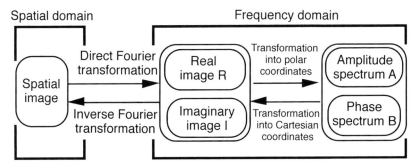

FIGURE 5.32 A schematic summarizing the transformation of a spatial image to the frequency domain. Each transformation is reversible. It would be possible, therefore, to see the influence of a convolution, carried out in the frequency domain, on the spatial image.

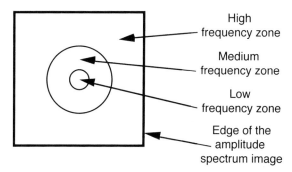

FIGURE 5.33 Rough limitations of the frequency zones in the image of the amplitude spectrum.

On the amplitude spectral image, the central zone characterizes the low frequencies and the outer zones characterize the high frequencies. Between these two zones, we find the average frequencies. It is in this zone that we find star-like objects.

The phase spectral image characterizes the position of an intensity variation on the spatial image. Let us take a single star at $(x_1;y_1)$ coordinates on spatial image field number 1. Let us now take the same star, present at coordinates $(x_2;y_2)$ on image number 2. We assume that, in the frequency domain, spatial images 1 and 2 will be transformed into the exact same amplitude spectral image. Only the phase spectral image will be different. We can assume that, roughly, the effect of a translation on the spatial image only affects the phase spectrum. We will see the advantage of this property in the study of high angular resolution of double stars.

Note that the currently used Fourier transformation algorithms are of the

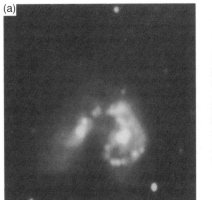 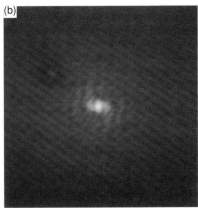

FIGURE 5.34 (a) The spatial image of the double galaxy NGC4038. (b) The image of amplitude spectrum shows three important regions: the central zones correspond to the low frequencies and contain information on the galaxies' halo structure. The streaked intermediate zone contains information on the field stars and on the HII regions of the galaxies. The outer zone, almost zero, contains information on the visible noise of the sky background. Image: B. David, A. Klotz and G. Sautot, Association T60.

FFT type: Fast Fourier Transform. They are, therefore, fast enough, but have the drawback of only working with square images whose dimension of a single side is a factor of two (generally, 64, 128 or 256 pixels).

Some FFT algorithms provide images from the frequency domain with high frequencies in the center and low frequencies on the sides. To return to the description of figure 5.33, a cross-permutation of the four image quadrants must be executed before applying the filters.

The Fourier transformation of an image is generally obtained by an optimized method called the FFT (fast Fourier transform). Hence, we generate two frequency images, R and I (real and imaginary) in Cartesian coordinates or A and P (amplitude and phase) in polar coordinates. These images allow us to separate images' different structures (stars, galaxies, noise, etc).

It is not only Fourier transformations which can be used to filter images in the frequency domain. Cosine and sine transformations provide a single real image. The null frequency is not placed in the transformation's center but, rather, in one of its corners. There are also transformations which generate images in the form of matrix blocks. In this family, we will cite the Slant, Hadamard and Haar transformations. Finally, the Karhunen–Loeve transformation relies on the development of a satisfying transformation with specific algebraic properties. These transformations are rarely used in astronomy

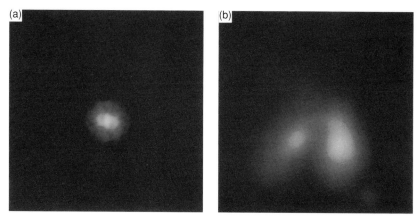

FIGURE 5.35 (a) The image of the amplitude spectrum from figure 5.34 has undergone a low-pass filter effect. The cut-off threshold has been chosen in order to suppress the high and intermediate frequencies. (b) After an inverse Fourier transformation, the resulting spatial image seems very smooth. Thus, we isolate galaxy haloes. Images: B. David, A. Klotz and G. Sautot. Association T60.

because of the difficulty physically interpreting the transformed image. However, they provide powerful data compression algorithms.

Convolutions in Fourier space Convolutions done in the frequency domain only concern amplitude spectral image modifications. If the phase spectrum is touched, the frequencies will be shifted and we will arrive at an image that no longer resembles anything, which is not the desired goal for a convolution.

The terms 'high-pass' and 'low-pass' have significance when convolutions are used in the frequency domain. According to figure 5.33, to create a high-pass filter, it suffices to set the pixels located in the low-frequency zone to zero. To create a low-pass filter, it suffices to set the pixels in the high-frequency zone to zero, which results in multiplying the amplitude spectral image by a filter whose central pixels are equal to 1 and fall to 0 toward the sides.

Experience shows that if the filter has a rapid transition from 1 to 0 we create oscillation patterns on the spatial image created from the inverse Fourier transformation. Hence, we must use filters which progressively pass from the coefficient multiplicator 1 toward 0 according to a decreasing law which will depend on the method used. Generally, we use Butterworth filters or exponential filters.

In practice, we execute the Fourier transformation directly from the spatial image to obtain R and I. Then, we apply a polar coordinate transformation in order to obtain A and P. Next, we multiply the A image by a high-pass or low-pass filter of a suitable form. All that is left to do is transform image A and P into

Cartesian coordinates and apply an inverse Fourier transformation to recover the convolved spatial image.

Frequency images are simply convolved by a point to point multiplication operation, which is faster than the same convolution in the spatial domain. We begin, thus, selectively suppressing certain undesirable image structures (noise or interference, for example).

There are not only high-pass and low-pass filters in the frequency domain, we also find pass-band filters, very useful for eliminating an interference signal or for extracting well defined frequency structures. A nice application is multi-frequency analysis in wavelets. In practice, we operate on the frequency images with filters that follow the Morlet wavelet function, with parameters chosen in relation to the frequency we want to analyze.

5.6 Image restoration

The angular diameter of stars is so small (about 0".05 for the largest) that no amateur telescope can measure their diameter. This leads us to thinking that each star, theoretically, occupies only a single pixel on the CCD array. But this is never the case since the 'atmospheric+optic' combination induces an unsharpness that spreads the star's image. Hence, we can imagine that each observed star on the CCD field acts like a single pixel which is 'fitted' with a strange hat that spreads out over several pixels. Seriously though, we can also imagine the problem by considering that each star only occupies one pixel which is convolved with a low-pass filter as in figure 5.25.

Figure 5.36 shows that the ideal image can be found from the observed image deconvolved by the function rendering it unsharp. This unsharp function is the instrumental function, also called the PSF, for point spread function. We must, therefore, know the PSF of an image to deconvolve it.

In practice, it suffices to isolate a star on the observed image and tell the software that it is the PSF itself.

Image restoration consists of deconvolving the observed image by its PSF (point spreading function).

There are numerous image restoration algorithms. Their common characteristic is their complexity! Only the best image processing programs allow restoration. As in the case with convolution, deconvolution can be carried out in either the spatial or frequency domains.

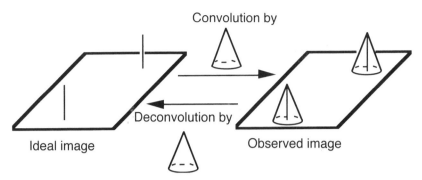

FIGURE 5.36 On the left, two point source stellar images can be seen in perspective. On the right, the same image, as observed in reality, is the result of a convolution. The image restoration consists of an inverse operation, that is, a deconvolution.

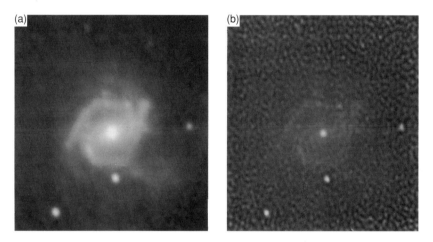

FIGURE 5.37 (a) A preprocessed image of Messier 61 obtained by compositing five exposures of 5 minutes with an ST4 CCD camera at the focus of a 180 mm $F/D=6$ telescope. (b) After the application of inverse filtering, the noise is revealed to the detriment of the details. CCD image: Jean Balcaen, Christophe Boussin and Vincent Letillois, Club Astronomique Science et Culture en Champagne-Ardennes.

We have seen in section 5.5.2 that the Fourier transformation of a spatial image leads to two images A and P, respectively called spectral amplitude and phase spectrum. We have just seen that it is necessary to extract an image from the PSF by isolating a star from the spatial image to deconvolve it. We must also, therefore, carry out the Fourier transformation of the PSF, whose result is called the modulation transfer function, or MTF.

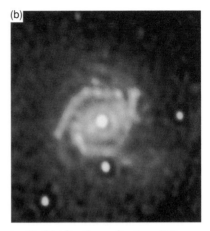

FIGURE 5.38 (a) A preprocessed image of Messier 61. (b) After the application of Wiener filtering, the galaxy's HII regions are revealed, but also notice the appearance of dark unaesthetic rings around the stars. CCD image: Jean Balcaen, Christophe Boussin and Vincent Letillois, Club Astronomique SCCA.

5.6.1 *Direct algorithms*

We have seen that to convolve the amplitude spectral image by a filter, it suffices to multiply the spectrum by the filter. Logically, the deconvolution operation consists of dividing spectral amplitude A by the MFT's spectral amplitude. It is, therefore, a rather quick operation, known as inverse filtering.

The major drawback of inverse filtering is that it strongly amplifies high frequency noise which sometimes renders the deconvolution results useless. One method, derived from the first, was developed by Wiener and keeps count of noise presence by the introduction of either the spectral amplitude of the image noise or, in its simplified version, a free parameter. If the parameter is equal to zero, we are left with inverse filtering. The main danger of this method is that we can reveal details that are actually only noise!

The geometric average filter is an even more evolved version of the Wiener filter since two supplementary parameters are introduced. Its use needs a mastery of noise amplification problems in order to conveniently choose the best parameter values. Let us also cite the blind deconvolution algorithm which allows the determination of the deconvolved image and the PSF at once. The calculation is done from any PSF image. The deconvolved image is calculated by the first PSF with the Wiener algorithm. The PSF image, therefore, is deconvolved with the previously calculated image and the observed image by once again using the Wiener algorithm. The observed image is deconvolved by the

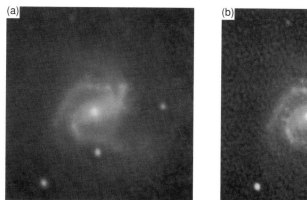

FIGURE 5.39 (a) A preprocessed image of Messier 61. After a Van Cittert algorithm restoration, a very fine image is obtained, which, unfortunately, increases the sky background's noise a little too much. CCD image: Jean Balcaen, Christophe Boussin and Vincent Letillois, Club Astronomique SCCA.

new PSF and so on until we obtain a stationary solution. Careful, this algorithm converges as slowly as the starting PSF is distanced from the real PSF and diverges in many cases.

5.6.2 *The Van Cittert algorithm*

Until now, we have covered direct methods since deconvolution is done in a single step. There are also, what are called, recurrent methods since they call for reiterative processes. The most famous of these is the Van Cittert method, invented in 1931. The Van Cittert method was improved by Jansson who introduced noise absorption parameters. The iterative methods restore the image in a much softer way than the Wiener algorithms. The iterative process is stopped the moment we judge the noise augmentation to be inhibiting. We often obtain an acceptable result after a dozen iterations.

Let us briefly examine the Van Cittert based iterative restoration algorithm: $I_{(0)}$ is the image to deconvolve and PSF is the PSF of the $I_{(0)}$ image, hence, the calculated image iteration $(k+1)$ will be $I_{(k+1)}$ defined by:

$$I_{(k+1)}(x,y) = I_{(k)}(x,y) + a(x,y)[I_{(0)}(x,y) - (I_{(k)}(x,y) \text{ convolved with the PSF})].$$

Notice the deconvolution is done in the spatial domain. For the Van Cittert algorithm, $a(x,y) = 1$. In the case of Jansson's algorithm, the relaxation coefficient

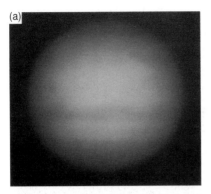 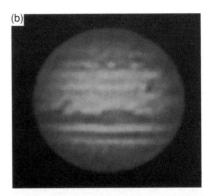

FIGURE 5.40 (a) A raw image of Jupiter obtained with a CCD camera focused at $F/D=19$ of a 400 mm diameter telescope. (b) After a Van Cittert algorithm restoration, a very good image is obtained, completely compatible with what would be obtained with an unsharp masking operation. CCD image: Gino Faroni.

$a(x,y)$ is quite complicated. It depends on an arbitrary constant chosen at the beginning equal to 0.1.

There are various other methods which only differ by the algorithm calculation $a(x,y)$. Note an important point: all of these methods require that not a single image $I_{(0)}$ pixel be negative. Therefore, use a point to point truncating operation for the negative pixels to return them to zero before restoring the image. The Van Cittert and Jansson algorithms, previously described, also exist in a version that can be applied to the image to deconvolve in the frequency domain. We prefer passing into the frequency domain when the PSF is large or complicated.

5.6.3 *The Lucy–Richardson algorithm*

Repopularized following the Hubble Space Telescope's disappointments, the Lucy–Richardson method (which dates from 1974) is also an iterating method which seems to give better results than the Van Cittert method for deep sky images. Unfortunately, both of these methods are not linear, therefore, the photometry is lost. The Lucy–Richardson method consists of the following operations, while keeping the same form as the Van Cittert algorithm:

$$I_{(k+1)}(x,y)=I_{(k)}(x,y)\,[\{I_{(0)}(x,y)/(I_{(k)}(x,y)\text{*PSF})\}\text{*PSF}]$$

The *PSF symbol signifies that we must convolve with the PSF.

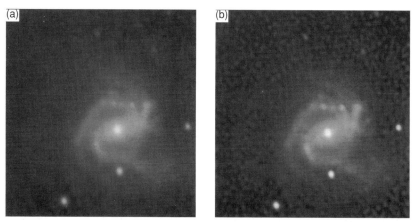

FIGURE 5.41 (a) A preprocessed image of Messier 61. (b) After a Lucy–Richardson algorithm restoration, a very soft image is obtained in which the sky background noise is not
as high as in the case of the Van Citter transdimension. CCD image: Jean Balcaen,
Christophe Boussin and Vincent Letillois, Club Astronomique SCCA.

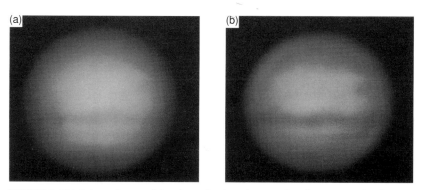

FIGURE 5.42 (a) A raw image of the planet Jupiter. (b) Note that the Lucy–Richardson
algorithm is not very appropriate to the increasing of planetary contrast. CCD image:
Gino Farroni.

5.6.4 *The maximum entropy algorithm*

A pixel's entropy is calculated by multiplying its value by its logarithm. An
image's entropy number is the sum of all of the pixels' entropies.

Methods, called MEM, for Maximum Entropy Methods, are based on the
idea that the restored image must satisfy two principle conditions: the image
must have the largest entropy possible and the convolved image with the PSF
must resemble the observed image as closely as possible.

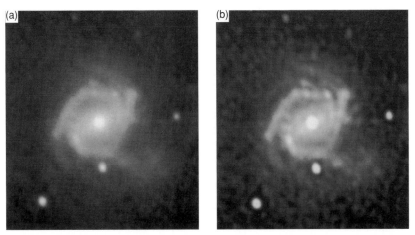

FIGURE 5.43 (a) A preprocessed image of Messier 61. (b) After a maximum entropy algorithm restoration, the HII regions are clearly revealed while the appearance of dark rings around stars is minimized. CCD image: J. Balcaen, C. Boussin and V. Letillois, Club Astronomique SCCA.

The deconvolved image's entropy is denoted S and the difference between the deconvoluted image and the observed PSF convoluted image is denoted C. The maximum entropy condition consists of maximizing the Lagrangian function $Q = (S - \lambda C)$. In practice, we seek to cancel this function's derivative. The resolution of this equation comes to the Gull and Daniell algorithm:

$$I_{(k+1)}(x,y) = \text{constant} \times \exp\left[-\lambda\{(I_{(k)}(x,y)^*\text{PSF} - I_{(0)}(x,y))^*\text{PSF}\}\right]$$

The exponential shows that the first iterations restore the image's large values. After a certain number of iterations, we see the image details, previously immersed in the noise appearing: this is where the algorithm becomes spectacularly powerful. Unfortunately, the exponential also introduces instabilities which often cause the algorithm to diverge, especially when the Lagrange λ multiplicator coefficient value is large.

In practice, we use rather more sophisticated maximum entropy algorithms where the other constraints are added to the desired image. The Q function is improved but we arrive, thus, at very complicated algorithms which are demanding of CPU time. These methods, however, are often used to synthesize radio astronomy images.

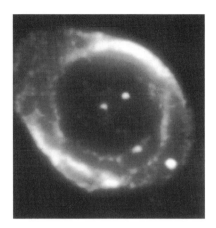

FIGURE 5.44 The Messier 57 nebula, obtained in a 10 minute exposure at the focus of an $F/D=23$, 60 cm Cassegrain telescope at the Pic du Midi Observatory, is restored using the hole algorithm associated with the wavelets algorithm. Image and treatment: Cyril Cavadore and Frederic Deladerrière, Association T60.

5.6.5 Diverse algorithms

We could write an entire encyclopedia on all the restoration methods, their efficiency and limitations. We have described the most important, but the bulk of restoration methods is filled with algorithms from which we will cite: the CLEAN algorithm, the minimum information method (MIM), the iterative bloc restoration, the Gerchberg method and the hole associated with wavelets algorithm.

5.6.6 Application areas for different methods

In all cases, we look to execute a Wiener deconvolution since the result is quite fast (we will make several attempts with reduction coefficients of different noise frequencies). It is also the algorithm that will best keep the photometry of the image. Then, we can try improving the result by using iterative methods.

In the case of low noise images (planets, the sun, etc.) we use a Van Cittert algorithm. In the case of deep sky images which have a weak signal to noise ratio (galaxies, star and nebula clusters), we use the Lucy–Richardson algorithm. Finally, in the case of extremely noisy images (faint nebula and galaxy extensions) we use the maximum entropy method.

Wavelet deconvolution can be used advantageously on images whose details to be revealed are all about the same size, this is the case of globular clusters and HII regions in certain galaxies.

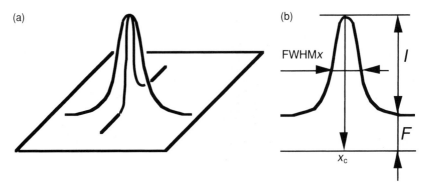

FIGURE 5.45 (a) A two dimensional Gaussian is usually used for modeling stars on CCD images. (b) Descriptions of the Gaussian function parameters on the x axis. On the y axis are the yc and FWHMy parameters.

5.7 Modeling

Modeling consists of synthesizing an image whose pixel intensity follows a mathematical law which best follows the reference image's pixel intensity. The model image, therefore, is free of noise and permits a simple morphological interpretation of observed objects.

The first step of such a treatment is choosing a parametered mathematical function which constitutes the model. The second step consists of extracting, from the reference image, the model's mathematical function parameter values. Finally, the third step consists of creating a synthetic image whose pixel intensity is only generated by the parameter values obtained during the preceding step.

5.7.1 *Stellar modeling*

Stellar modeling consists of synthesizing the stars' image. In general, the model's mathematical function is a Gaussian in which the parameters are: the center position (x_c, y_c), the width at half maximum on both axes (FWHMx and FWHMy), the light intensity (I) and the sky background's value (F) (see figure 5.45). The term FWHM signifies Full Width at Half-Maximum. The mathematical adjustment of the Gaussians on the stars is generally obtained by the method of least squares, which consists of minimizing the square of the residuals obtained from the subtraction of the Gaussian from the star.

If we model all of the stars on a CCD image, we can draw up a list which would contain, for each star, the six parameters describing each Gaussian (x_c, y_c,

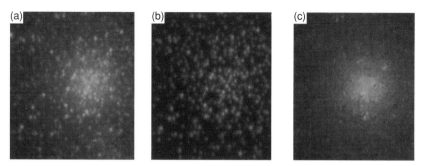

FIGURE 5.46 (a) An image of Messier 13. A modeling of the image's stars. (c) The subtraction of the cluster's stars reveals a continuous background due to the unresolved stars. Image: B. David, A. Klotz and G. Sautot, Association T60.

FWHMx, FWHMy, I, F). If the stellar field image is not too affected by the telescope's optical imperfections (aberrations, coma, etc, . . .), all of the stars must have the same full width at half-maximum FWHMx and FWHMy. For this, some algorithms constrain this condition and calculate a single average couple (FWHMx, FWHMy) which is applied to all of the analyzed field's stars.

The parameters (x_c, y_c, FWHMx, FWHMy, I, F) serve to calculate the Gaussian flux integral, hence, the relative magnitude. For this, we apply the following formulas:

$$\sigma x = 0.601 \text{FWHM}x, \qquad \text{Flux} = \pi \sigma x \sigma y I,$$
$$\sigma y = 0.601 \text{FWHM}y, \qquad \text{Relative magnitude} = -2.5 \log(\text{Flux})$$

The calculation of absolute magnitudes requires a photometric calibration described in chapter 6. The log notation signifies a base 10 logarithm.

There are no Gaussian functions that model stars well. The Moffat functions sometimes allow a better adjustment with the profiles of observed stars.

5.7.2 Isophotal modeling

Isophotal modeling creates a synthetic image constituted of a network of isophotes that best fit the isophote set on the initial image. The smoothing algorithms rely, generally, on the parameter search of a predefined mathematical function for each isophote. We can, therefore, produce the morphological analysis of different objects. The most current application consists of modeling elliptical galaxies by a purely elliptical network of isophotes. The subtraction of the model from the initial galaxy image reveals discrepancies from ellipticity of the galaxy.

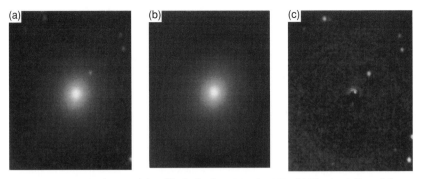

FIGURE 5.47 (a) An image of the elliptical galaxy Messier 105. (b) The modeling of the galaxy made up of a network of elliptical isophotes. (c) The subtraction of the model from the initial image allows the almost perfect erasing of the galaxy (with the exception of a few calculation artifacts in the center). Image: B. David, A. Klotz and G. Sautot, Association T60.

5.7.3 *Modeling of the sky background*

It is sometimes useful to synthesize the sky background in the form of a polynomial surface which best adjusts itself on a few dozen points chosen by the user. This can be used to correct non-uniformity faults that can persist after a photometric correction produced with a mediocre flat field. The polynomial layer can, therefore, function as a second flat field to be applied after the first. It is a method which allows the recovery of some desperate cases, but it is not suggested that it be used routinely; nothing replaces a good flat field!

Sky background modeling by a polynomial surface can be useful in removing a diffused object's contribution which prohibits the correct analysis of smaller sized objects overlapping it. It is, for example, the case of the small Messier 32 galaxy superimposing itself on the large Adromeda galaxy background.

5.8 Choosing image processing software

The choice of an image processing software must be guided by: the number of treatment and analysis functions, the power of the algorithms, the image's presentation on the screen, the user friendliness and the speed of execution. We have grouped, below, two lists of image treatment functions. The first list assembles those functions which are indispensable to all astronomical image processing software. The second list assembles the supplementary functions which allow amateurs to work like professionals!

Indispensable basic functions:

- Loading and saving images in the FITS format.
- displaying the image, with at least 64 levels of gray, while adjusting the high and low thresholds.
- Point to point cross-hair analysis, astronomical photometric windows analysis (centroid).
- Point to point treatment operations: addition or subtraction of a constant, multiplication of a constant, addition or subtraction of 2 images, division of 2 images.
- Geometric transformation operations: fractional pixel translations.
- High-pass and low-pass filter convolutions.

Evolved treatment functions:

- Large format image (at least 512/512 pixels) treatment.
- Display images with 64 levels of gray and 16 million colors for tri-color processes.
- Point to point treatment operations: production of a median image from a lot of images, dark optimization, RGB/HSI transformation, polarimetric coordinate transformation.
- Geometric transformation operations: mosaics, skeletization, anamorphic enlargement (Splines) or reduction, planisphere, polar projection.
- Frequential area convolutions.
- Wiener, Van Cittert, Lucy–Richardson and maximum entropy deconvolutions in the spatial or frequency domain.
- Stellar image synthesis modeling.

6 CCD applications in astronomy

6.1 The advantages of the CCD

A CCD's primary function is to produce images. At first, it appears to rival photography. Furthermore, it allows luminous flux measurements, and therefore rivals photometers.

It is in comparing its performance with existing detectors, photographic film in particular, that we can easily see in what areas the CCD will assert itself and what new areas can be opened up.

6.1.1 *Image quantification and linearity*

By its nature, the CCD image is digitized with a regular spatial sampling. Moreover, the digital value representing each image point (after the dark and flat field corrections) is proportional to the amount of light received. Hence, the image is directly usable in digital processing, which makes it accessible to powerful information extraction tools, described in chapter 5.

Digitally processing a photographic image is much less natural. We must first sample the image at regular intervals: a microdensitometer is placed in front of the film which measures its density over an area a few micrometers wide; but this measurement is not proportional to the quantity of light the film received and the area measured must be converted into the amount of light received by the film's standard response curve, for which there is no precise source. Furthermore the mechanism which moves the microdensitometer from one measurement zone to another on the film's surface must be accurate to within a micrometer, which is not easy to achieve. Thus, the application of numerical analysis to photographic images must pass through a tedious and expensive step, out of reach of the amateur astronomer.

Astronomical photometry needs measurements proportional to the quantity of light received by the instruments (a linear response), hence the inaccuracies of photographic photometry. Photometers are linear devices but are generally single channel: a single sky point is measured at a time. The CCD combines the photometer's linearity with the image's surface information. Hence, it becomes a valuable photometric tool.

Astrometry is another important sky image analysis discipline. For over a century, surveys of astronomical positions were produced by measuring the

images' positions on a photographic film or plate. This is tedious work, requiring good equipment, adroitness, and patience (in astronomical observations
during the 1900s, this work was reserved for women, who were supposed to be
more meticulous than men!). Now, the measuring problem is, of course,
resolved by the CCD: its pixel decomposition immediately gives access to the
coordinates of all objects present on the image.

6.1.2 *Detectivity*

Thanks to its excellent quantum efficiency and the absence of thresholds, the
CCD is a detector with a superior detectivity to film. In astronomy, a detector's
detectivity is a fundamental parameter. A large detectivity allows:

- The observation of faint objects.
- The use of narrow band filters in order to select spectral information.
- The use of long focal lengths, which give better sampling of the spatial
  image.
- The use of short integration times, which are fundamental for limiting
  the influence of atmospheric turbulence, but also permit us to better use
  observation time.

It can be seen that the CCD has a decisive advantage in all areas of observation,
from the faintest objects to the planets and even the Sun.

6.1.3 *Spectral sensitivity*

Compared to photographic film, the CCD covers a wavelength range twice as
wide: from 400 nm to 1 μm, while film, certainly sensitive in the near ultraviolet short of 400 nm stops at 650 or 700 nm, at the red edge of the visible. Hence,
the CCD opens new perspectives in imagery in the area of 700 nm–1 μm, which
was not previously covered by lower-performance infrared films.

6.1.4 *Real time work*

While photographic film needs to be developed, the CCD allows the image to be
displayed a few seconds after the exposure. Besides the operator's satisfaction,
and the reassurance and security with regard to possibly incorrect handling, the
real time aspect is an important advantage of the CCD in several types of
observations:

- At high resolution, with short exposure times, one method of limiting the influence of atmospheric turbulence, which has a random character, consists in selecting a few images – the best ones – from a large number of images. The CCD enables this selection to be made in real time, therefore avoiding the development (and wasting) of dozens of meters of film that the photographer disposes of after sorting through them.

- In the case of the discovery of a supernova, asteroid, comet, etc. the observer can immediately produce or request confirmatory images and raise the alert that same night.

6.1.5 *A restriction: the wide field*

Alongside the avalanche of advantages over photographic film, we must also state the CCD's main drawback, which is a very restricted field, with a comparable resolution, because of the size of the CCD arrays actually available to amateur astronomers.

However, an examination of various objects that are of interest to astronomers shows that most of them can be covered by the field the CCD offers. To obtain a complete image of objects that are too large, it is possible to produce a mosaic of several images, at the cost of the time devoted to acquisitions for as long as there are images to produce and the more complex processing required.

The main handicap of the field's size is apparent in sky surveys, where the number of square degrees covered per unit time is directly proportional to the efficiency of searching for new objects.

Because of its great detectivity, extended spectral sensitivity, array structure, linearity, analysis and image processing possibilities, and its real time aspect, the CCD is a particularly good detector, especially for imaging, photometric, and spectrographic tasks.

The main drawback of the CCD is its small size, but this handicap is only manifested on a limited number of observable subjects.

6.2 High resolution work

A celestial object's CCD image has as good resolution as the telescope's diameter is large, as its focal length is long, and as the camera's pixel dimension is small. In terms of sampling (the number of arcseconds per pixel), high resolution implies values in the order of 0.5–0.2″ per pixel. With a CCD camera whose pixels are 15 μm in size, a 0.2″ sampling per pixel implies having a focal length of 15 meters. For amateur telescopes, such a focal length is generally not available except with a multiplying lens (a Barlow lens or eyepiece). Unfortunately, resolution is limited by atmospheric turbulence. One tries to suppress the effect of turbulence by carrying out the shortest exposure possible. Finally, even with short exposure times, all images are not affected in the same way because of the atmosphere's random agitation. One therefore tries to directly select the best images over shorter intervals, which requires the use of a CCD camera whose read speed is as fast as possible.

The acquisition of images is done in 'automatic sequenced' mode: the camera provides an image, it is displayed on the screen with the proper display thresholds, the exposure time is refined, and then a new exposure is requested. Once it is displayed, we judge its quality by comparing it to the previous image. If it is satisfactory we save it; if not, we take another exposure. And so on – we can easily see 300 images per hour. Direct selection, such as we have just described, is necessary since it is impossible to store all of the images onto the hard drive. On average, only one image out of about 20 is saved.

Several dozen raw images can be stored per night, which shows that it is sensible to equip oneself with a high-capacity hard disk (over 100 megabytes). In order to reduce the space taken up by images and reduce the CCD's read time, one tries to window the image, during the acquisition, to eliminate the image's border pixels which do not contain any information. However, a sufficient margin should be left around a planet (at least 15 pixels) to avoid edge effects, caused by some forms of image processing, so as not to disturb the zone occupied by the observed object. One must remember to window the images used for the preprocessing (precharge, dark and flat fields) in the same way!

High resolution CCD observation requires image acquisition with high sampling but a short enough exposure time to 'freeze' the atmospheric turbulence. The CCD camera's sequenced automatic acquisition mode allows the selection of the best images during image capturing. When using large CCD arrays, we try to window an object's image at the moment of acquisition in order to reduce the CCD's read time.

6.2.1 *Planetary surfaces*

The observation of planetary surfaces is a privileged activity area for amateur astronomers, whether it is done visually or photographically. The CCD opens up a new dimension.

Image acquisition We look to use the longest focal length possible with the largest possible number of pixels per arcsecond. But the use of a long focal length has two limitations:

- Firstly, the planet must be contained on the CCD's surface. The extreme case is that of fitting the largest planet, Jupiter, with its 44″ at opposition, in the smallest CCD used by amateur astronomers: the TC211 with its 165 pixels per side. By making a ten pixel margin on either side of the planet, we see that the sampling is 0.3″ per pixel, which is completely acceptable in terms of image quality for most amateur sites; the corresponding focal length, therefore, is 11 meters. The observation of other planets or the use of larger CCD arrays lead to perfectly acceptable situations in terms of field availability. Of course, the Moon is another problem: sufficiently sampled to maintain a high resolution, it would not fit in its entirety on any array used today by amateur astronomers (the Moon is close to 2000″ in diameter).

- Next, a long focal length gives an extended image which is not very bright, for which we increase the exposure time to obtain a large signal. But the exposure time is limited to a few tenths of a second to reduce the effects of atmospheric turbulence. If the telescope has a small diameter, obtaining a larger signal will be incompatible with a large sampling, despite the CCD's high sensitivity. The problem occurs most often when the use of filters reduces the luminous flux; we must therefore find a compromise in order to maintain a short exposure time: a high sampling level, leading to a relatively weak signal, favors small contrasted details; a low sampling, leading to an optimum filling of pixels, allows better detection of lower contrast details.

Remember that it is impossible to capture images of the Moon with a full frame camera not equipped with a shutter.

Image selection, keeping only those that are least altered by turbulence, is essential. Some processing, such as the three-color process, needs the use of several selected images, obtained in a short enough interval that the planet's rotation is negligible on all of the images. It is therefore necessary to take these images at the highest possible rate and efficiently make selections. The camera's read time must be as short as possible and its software should have display functions that allow the operator to select images, quickly and surely.

It would be useful if the acquisition software quickly heightened the contrast before displaying the images in a sequenced automatic mode. One could then judge more easily the quality and make a more objective selection. Meanwhile, it is imperative that the software save the raw image on disk and not the processed image (in order to be able to carry out the flat field corrections and further optimize the contrast enhancement method).

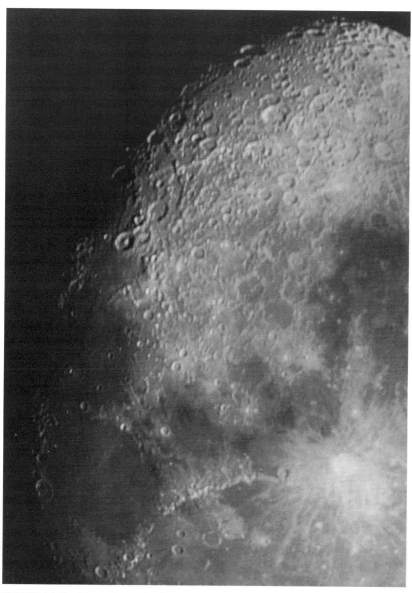

FIGURE 6.1 The Moon. A mosaic of two CCD images produced with an Alpha 500 camera at the focus of a 300 mm diameter telescope with a focal length of 1800 mm. Exposure: 0.01 seconds. Each elementary image's field is 17 arcminutes and the sampling is 2 arcseconds per pixel. CCD image: C. Cavadore, P. Martinez, P. Montferran, L. Pinna.

Basic image processing The preprocessing step has as its main goal correcting the image of high-frequency defects: dust and inter-pixel response variations. Beware: we must use a flat field produced with the same filter as the image. Because of the very short exposure times used, we often prefer not to dark correct so as not to add read noise (we simply remove the precharge image from the raw image).

The second step consists in improving the local contrast. There are two methods for this and both can be used independently or one after the other, depending on the case:

- Unsharp masking (section 5.5.1), which greatly increases the local contrast of an image, even if it has a strong global contrast (in the absence of this strong global contrast, it suffices to increase the image's dynamic range to make the details appear without saturation).
- Deconvolution functions, like the Van Cittert iterative restoration allow the sharpness of details to be improved.

The preprocessing and contrast increasing steps are applied to all nighttime images. It is better to use image processing software that allows the creation of a large preprocessing macrofunction which could be automatically applied to the dozens of nighttime images. To facilitate this treatment mode, we suggest that the raw images are saved under a generic name, such as JUP, for example, followed by the chronological save number. For example, over the course of the night, if we save 348 images on the hard disk, the file associated with the first image is named JUP1 and that of the last image is JUP348. Normally, the date and time of the file saved onto the disk can be read by reading the file's name (the DIR command in DOS). If the images are saved in the FITS format, ensure that the header contains the date and time of the exposure.

Use of data Amateur CCD images contain enough information to allow detailed studies of Mars (climatological phenomenon research, such as sand storms and fog) and Jupiter (cloud mass evolution and displacement).

We can either work on isolated images, or synthesize a cartographic projection (section 5.4.3) of simple cylindrical type or USGS, then produce a complete planisphere by mosaic using several images taken at different moments during the planet's rotation.

Since Jupiter turns once on its axis in about ten hours, it is difficult to cover the entire planisphere in a single night, except during winter oppositions. In general, this can be done by combining the images obtained over the course of two or three consecutive nights.

Mars's case is more difficult: this planet turns once on its axis in about 24 hours; because of this, it shows the same side to the observer over several consecutive nights. The planisphere's mapping requires observations spread out

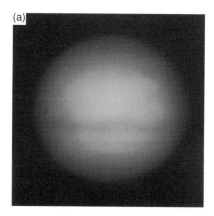

FIGURE 6.2 (a) The raw image of Jupiter has very low contrast. (b) After the unsharp masking treatment, details in the planet's band begin to appear. (c) After the unsharp masking treatment, the image was restored with a single iteration of the Van Cittert algorithm in order to considerably augment the contrast of details. Image: G. Farroni.

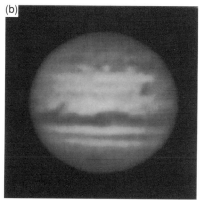

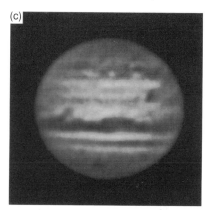

over a long period, at least two or three weeks and it is not uncommon to see the planet's appearance change during this time (with melting of the polar caps, for example!). A better solution would be the cooperation between several observers spread in longitude on Earth: through relaying, a continuous 24 hour observation of Mars can therefore be guaranteed.

To map out a complete planisphere of a planet, one can proceed in stages in the following order:

- For all the preprocessed images, increase the contrast by the unsharp masking method and if necessary, correct for anamorphism due to the camera's rectangular pixels (see figure 6).

- Select the best images of the preprocessed series. The selected images are regularly spaced in terms of the central meridian longitude. In general, choose an image at every 20–30°, which corresponds to about 15 images for a complete rotation.

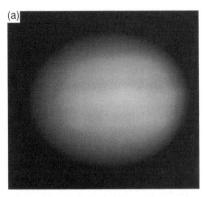 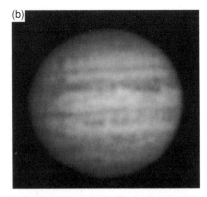

FIGURE 6.3 (a) The raw image directly output from a LYNXX camera. The camera's rectangular pixels produce an excessive flattening of the planet. (b) The image has been treated with unsharp masking and has undergone anamorphous correction by enlargement of the vertical axis in order to remove the flattening due to the rectangular pixels. Image: G. Farroni.

FIGURE 6.4 A coordinate grid is adjusted, by the operator, on the image of Jupiter in order to prepare the image processing software for carrying out the cartographic projection. Image: G. Farroni.

- Search for the planispheric projection parameters for each of the selected images. These parameters can be finely adjusted by superimposing the coordinate grid on the images.

Synthesize a planispheric image for each of the selected images. It is theoretically feasible to cover a longitudinal spread of $\pm 90°$ from one side of the central meridian to the other. It is strongly suggested that the projection be limited to $\pm 60°$, while the planet is at opposition, because of the fact that the planet's edge details are seen at a steep angle. Once the planet shows a phase greater than $5°$, this interval is limited on the side of the phase. The latitude projection is also limited, for the same reasons, $\pm 60°$.

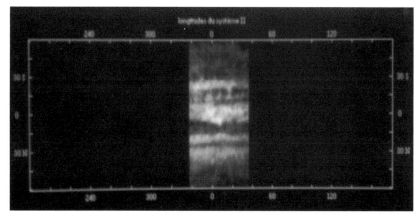

FIGURE 6.5 A cylindrical cartographic projection of the image of figure 6.4. The projected zone's extent has been purposely limited in order to avoid the appearance of asymmetrical lighting faults due to a 10° phase. Image: G. Farroni.

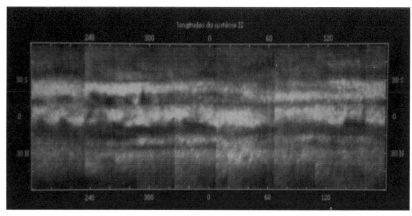

FIGURE 6.6 A cartographic projection showing the entirety of Jupiter's surface. This image is a result of the mosaic of 17 partial planispheres taken between 25 June 1993 at 20hUT and 28 June 1993 at 21hUT. Note that these images were obtained three months after opposition and that the planet was only visible during a 2 hour period per night. This planisphere consists of the most mediocre quality images we can make! Image: G. Farroni, $F/D = 19$, 400 mm telescope. Image treatment: Alain Klotz.

- Produce the complete planisphere by carrying out a mosaic of the partial planispheres obtained during the previous step. Experience shows, in the overlapping zones between partial planispheres, that it is preferable to attribute, to the total planisphere, those that have the maximum value.

FIGURE 6.7 From the planisphere shown in figure 6.6 an oblique orthographic projection can be carried out allowing us to see the planet from any point in space. Image: G. Farroni. Image treatment: A. Klotz.

From a planisphere resulting from a cylindrical coordinate projection, it is very simple to measure the coordinates from one detail or another and calculate their movements from several planispheres obtained on different dates.

Image processing functions, such as deconvolutions and unsharp masking, allow the CCD to detect planetary details not easily accessible in photography. It is also possible to produce planispheres.

Filtering of planetary images The study of planetary surfaces often implies the use of filtered images: either to produce three-color maps or to study phenomena or the movements of matter which is easier to identify at certain wavelengths.

To produce three-color images the BGR photometric bands are used in the standard way. Switching from one filter to another must be done, as far as possible, without having to refocus. For this, the filter combinations for each color must be of the same glass thickness. Each three-color process (use of three filters) must be done in the minimum time (less than 3 minutes for Jupiter, 15 minutes for Mars) in order not to show the effect of a planet's rotation. The exposure times will have to be adjusted to have the maximum dynamic range while not saturating. Note, however, that the blue images must be exposed longer than the others, mainly because of the CCD's weaker sensitivity to shorter wavelengths.

The planet Venus does not show any notable coloration. We are looking, rather, to reveal cloudy structures that often appear in the blue image. For this, a two-color image is taken in the B and G bands. During the processing, the G image is used as an unsharp mask. We begin by recentering the G image in relation to B, then divide image B by image G. The result is multiplied by 10, for example, then we add the result to the original blue image.

Concerning Earth-like planet (the Moon, Mars) soil composition studies, we can identify the presence of a few minerals by analyzing the λ_1/λ_2 image resulting from the division of two images centered on the λ_1 and λ_2 wavelengths. The filters needed must have the narrowest possible pass-bands. For the Moon, we suggest capturing images only during a full moon since we must avoid phase effects as much as possible. Here are two interesting cases to analyze:

- $\lambda_1 = 400$ nm and $\lambda_2 = 560$ nm. If the pixel value of the 400/500 image is lower or equal to one, the titanium dioxide abundance (TiO_2) is lower than 2%. For a value of 1.10 the abundance is about 10%. Between 1.00 and 1.10 we can roughly linearly interpolate the TiO_2 abundance.

- $\lambda_1 = 700$ nm and $\lambda_2 = 900$ nm. The pixel value of the 700/900 image is as greater than 1 as the olivine or pyroxene abundance is high. Indeed, we can increase the contrast between the two types with the 1100/900 image: the presence of pyroxene is identified for pixel values above unity.

The Moon has a few interesting phenomena to search for with the CCD, for instance, TLPs (transient lunar phenomena). These are surface zones about 15 km in diameter which become brighter than usual for about 20 minutes on average. There is little quantitative data for TLPs and CCDs should easily enable them to be measured. The zones to look out for are the edges of seas and certain craters (Aristarchus and Alphonsus), especially when the Moon is close to perigee. Concerning the use of filters, it seems that the TLPs' luminous intensity is stronger in red light. The image resulting from the R/B difference could therefore be used to locate TLPs more easily.

For the giant planets, it could be interesting to study the methane distribution. Hence, one would use a pass-band filter centered on 886 nm with a width of 5 nm. The exposure times are very long, in general between 30 seconds and 2 minutes. This area is reserved for very good sites and telescopes which track well.

6.2.2 *Double stars*

A large proportion of stars form multiple systems. At the amateur level, the observation of double stars consists in measuring the difference and position of one of the components in relation to the other. The measurement of cleanly separated stellar doubles is not linked to the high-resolution field but the field of astrometry (section 6.5). We will now study the case of double stars, typically separated by at least 10 arcseconds. Before outlining the observation method, let us recall, historically, the different measurement methods.

Usually, the measurement of double stars is done visually with the help of a special eyepiece equipped with a micrometer. For the experienced observer, with the help of a large instrument, the visual method allows the measurement of differences as small as a fraction of an arcsecond on stars of the 10th magnitude, in only a few seconds! Among the masters of the subject is Paul Couteau, an astronomer at the Nice observatory.

Thanks to the regular geometric arrangement of its pixels, a CCD camera allow the precise measurement of the spatial configuration of the field's stars, that is, the determination of their angular separation and their position angle. In general, the measurement done with a CCD camera takes much more time than a visual one because of the time taken to process the images. Nevertheless, the CCD has the advantage of providing precise results that are completely objective and gives, in certain cases, a measurement of the star's magnitude.

Image acquisition The image must be as oversampled as possible. We must therefore enlarge the image with a Barlow lens or an eyepiece. However, note that we lose about 1.5 in magnitude limit when we enlarge by 2 times. Also, we try to 'freeze' the turbulence by having the shortest exposure time possible. We must not hesitate to use the shortest exposure time available on the camera. With such exposure times, we strongly suggest having a mechanical shutter if a full frame array is being used (if the field contains only the double star, we could use the half-frame mode without a shutter). The need for a short exposure time and the desire to observe the faintest possible stars requires a camera which has the best possible read noise.

Concerning the acquisition session itself one proceeds in the same way as with planetary imaging. The best images are selected instantly. In general, at least 10 images are kept for each binary. Besides the double star images, we suggest taking images that allow orientation and field scale calibration. Be careful, however, for the orientation of the field of an instrument installed on an azimuthal mounting changes with the direction of targeting. Therefore, avoid observing double stars with such mountings.

The field's orientation is easy to determine since it suffices to let the star pass while stopping the telescope's tracking. Notice that the star does not follow a perfectly linear path, but displays some oscillations characteristic of turbulence. The tracking of two stars separated by 10 arcseconds, makes it possible to judge the size of the field of coherence, that is, the field width in which the stars are affected the same way by the turbulence.

The image scale, that is, the sampling, can be measured with the help of a known double star separated by half the field. We could even determine the field's orientation by this method, but the orbital motion of the components is often too poorly known to permit sufficient precision for the orientation angle.

FIGURE 6.8 Tracking the double star 100 Herculis. The tracking follows the east–west direction, allowing the calibration of the CCD field's orientation. Notice that the trails are affected by the same small oscillation. This signifies that both stars are in the same coherent field of turbulence.

FIGURE 6.9 Composite of two images of the double star 100 Herculis (separation of 14″.1). The stars' parallelogram allows the calculation of the orientation and scale of the field. The direction of the parallelogram's longer sides' determines the east–west orientation. The length of the shorter sides represents the double star's separation, with which we determine the sampling (scale) Image: G. Faroni, A. Klotz and G. Morlet.

A method developed by Guy Morlet in 1991 determines the field's orientation and scale straight away. For this, the image of a double star located near the CCD field's side is produced. Then, the stellar double is moved toward the opposite side by using the right ascension drive motor. A second image is therefore produced. Finally, the two images are added together. Hence, the resulting image shows four stars which simultaneously allow the measurement of the scale and the east–west orientation of the image with the help of the photometric centroid calculation.

Data reduction We can distinguish two types of images according to the separation of the binary. The simplest case concerns perfectly separated stars where we can measure, without ambiguity, the centroid position of each star. The parameter determination is done by measuring the positions of the centroids (see section 4.3.3).

Double star
image

Super position of
the shifted image

Autocorrelation
image

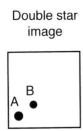
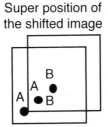
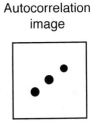

FIGURE 6.10 When the image of the double star A–B is moved by the amount of their separation, it is possible to superimpose the components A and B. This results in the appearance of a non-central peak in the autocorrelation image. The same shift value, on the initial image, in the opposite direction, produces in the autocorrelation image a secondary symmetrical peak with respect to the center.

The most complicated case obviously concerns stars that overlap one another. A 'separation' method consists in synthesizing an autocorrelation image of the stellar binary and suppressing the image's central peak.

The autocorrelation image is a spatial area image whose pixel intensity takes into account the initial image's degree of resemblance to itself, after displacement by a certain number of pixels. On the autocorrelation image, the number of shifted pixels is counted in relation to the center. For example, any image will resemble itself for a zero pixel shift; the autocorrelation image, therefore, always has its central pixel non-zero. The autocorrelation image of an image which has only a few stars will have its central pixel non-zero as well as two other pixels that are as far from the central pixel as the distance between the double star. The position angle of the autocorrelation image's non-central peaks will be identical to the position angle of the double star. In reality, the exact theory is a little more complex and it is possible to find other secondary peaks on the position angle axis.

The advantage of an autocorrelation image is that it is identical, regardless of the stellar binary's position on the raw image. Hence, we can add the autocorrelation image's intensity without worrying about the shift produced by the atmospheric turbulence between the raw images.

In practice, the autocorrelation image is obtained by executing a Fourier transformation (see section 5.5.2) on the spatial image. We transform into polar coordinates in order to have access to the amplitude and phase spectra. All of the phase image's pixels are zeroed, we transform the Cartesian coordinates, and then carry out the inverse Fourier transformation. Finally, cross-permute the spatial image's four quadrants (see section 5.4.1).

The autocorrelation image's pixel value is symmetrical in relation to the center. The image shows a central peak and secondary peaks situated in the

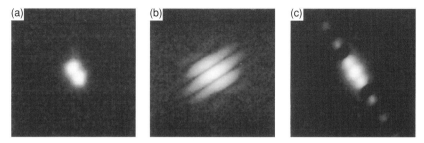

FIGURE 6.11 (a) The raw image of the double star 49 Serpentis (STF2021) whose $V=6.7$ and $V=6.9$ components are separated by 4".26 (1971 data). (b) The amplitude spectral image shows the fringe interference phenomenon. (c) The autocorrelation image shows a series of peaks aligned according to the position angle's direction. The separation value of the double is found by measuring the distance from the first secondary peaks to the central peak. Image: G. Farroni, A. Klotz and G. Morlet, LYNXX camera, T400, $F/D=19$; 6 seconds.

double star's position angle direction. In the case of stars of comparable brightness, there are many secondary peaks. Note that the distance separating the central peak from the closest two secondary peaks is the desired separation distance.

We have just seen how to measure the double system's separation and position angle at once. This method allows an additional subtlety: canceling the image from the phase spectrum makes the measurement independent of the stellar binary's original position on the raw image. Regardless of the binary's position on the raw image, the amplitude spectrum is identical. We can therefore composite a large series of short exposure images by simply adding their amplitude spectra. The autocorrelation image will be of better quality.

If the binary's separation to measure is less than a dozen pixels, it is often difficult to separate the closest secondary peaks from the central peak on the autocorrelation image. And if the stars from the couple are not close in magnitude, only the first secondary peaks appear, but since they are flooded in the central peak's halo, a measurement becomes impossible. This case is perfectly illustrated with the Alpha Herculis star whose components are $V=3.0$ and $V=6.1$ for a separation in the order of 4.9 arcseconds.

To avoid this last problem, it suffices to remove the central peak from the autocorrelation image. For this, we calculate the polar projection of the autocorrelation image centered on the central peak (see section 5.4.3). We reattribute, to all the pixels from a line with the same distance from the center, the median value of this line. We apply this principle for all the lines and then subtract this new image from the previous one. There must only remain details which are different from the central peak's circular symmetry.

FIGURE 6.12 (a) A raw image of the star Alpha Herculis (STF2140) whose components are at magnitude $V=3.0$ and $V=6.1$ and are separated by 4".9. (b) The amplitude spectral image. (c) The autocorrelation image only shows two faint secondary peaks, poorly separated from the central peak. Image G. Farroni, A. Klotz, and G. Morlet, LYNXX camera, T400, $F/D=19$; 0.01 seconds.

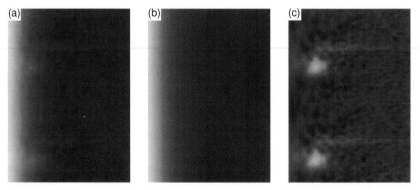

FIGURE 6.13 (a) The autocorrelation image of the star Alpha Herculis has undergone a polar coordinate transformation centered on the central peak (the θ angle axis is vertical and the distance axis is horizontal). (b) The image in (a) has been smoothed by a median filter applied on each of its columns. (c) The subtraction of these two images allows the extraction of the secondary peaks' components by erasing the central peak.

We return to Cartesian projection in order to find the autocorrelation image's initial geometry. We notice, then, that the central peak has disappeared. In return, the first secondary peaks are now very visible and allow the very precise calculation of the position angle and the couple separation with the help of the photocentroid.

Different digital image treatment functions allow the measuring of a double star's parameters, even when the two components' images are close and overlapping one into another.

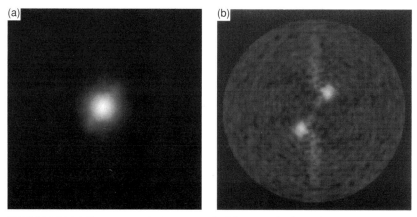

FIGURE 6.14 (a) The autocorrelation image of Alpha Herculis shows faint secondary peaks. (b) After the subtraction of the central peak (see figure 6.13) the double star's separation parameters can finally be precisely measured. Image: G. Farroni, A. Klotz and G. Morlet, LYNXX camera, T400, $F/D=19$.

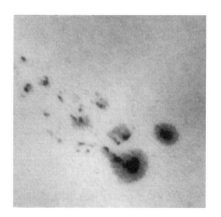

FIGURE 6.15 An image of the Sun obtained on 31 May 1993 with a LYNXX camera at the focus of a telescope equipped with a neutral filter. Exposure time: 0.02 second. This image was restored through 10 Van Cittert algorithm iterations. CCD image: Jean-Marie Llapaset.

6.2.3 *The Sun*

The Sun is one of the high resolution imaging applications. Unfortunately, most CCD camera first-time users are exclusively devoted to the night sky, because of the advantages the CCD has, due to its heightened sensitivity, while there is no lack of light in solar imaging. Consequently, not much solar CCD work has been produced as of yet.

Meanwhile, it is obvious the CCD will, in this area too, improve the work achieved to date by photography.

The CCD advantages for solar imagery are:

- Its extended spectral sensitivity. In particular, the CCD is very sensitive to the Hα to 656 nm wavelength, quite important in chromosphere and corona studies.

- Its large dynamic range. Some fine solar surface structures have little contrast: granulation, for example, and internal spot structures. A 16 bit CCD camera should reveal details inaccessible to photographic films.

- Its real time image selection possibilities, which avoids turbulence more easily. Remember that it is impossible to produce solar images with a full frame camera not equipped with a shutter.

6.3 Images and detection of faint objects

6.3.1 *The field of study*

Galaxies, nebulas, star clusters, comets, asteroids and, generally, objects not visible to the naked eye are classified as faint objects. Their common characteristic is that it is necessary to carry out long exposures, which requires the use of well cooled CCD cameras, preferably with an MPP technology array.

The CCD's major advantages in this application field is, its great sensitivity and its digital image processing possibilities, not to mention its extended spectral sensitivity covering, in particular, the Hα line, important for studying nebulae.

6.3.2 *Image acquisition*

In general, the camera is mounted at the prime focus of the instrument. We verify anyway that the sampling, already defined, allows us to read the resolution limited by the long exposure turbulence of the site. In general, we sample the images between 1 and 4 arcseconds per pixel, depending on the CCD's quality (for example, an Alpha 500 camera used with a 2 meter focal length gives a 2″ per pixel sampling). Meanwhile, some applications can lead to undersampling, in order to keep a considerable field despite the limited number of the CCD arrays' pixels; this is generally the case with optics which have a focal length smaller than 1 meter (for example, used with a 500 mm focal length, the Alpha 500 camera gives a 8″ per pixel sampling, but covers a field slightly greater than 1° wide).

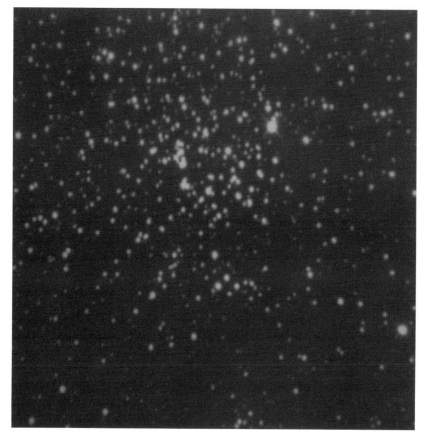

FIGURE 6.16 The open star cluster Messier 52 in Cassiopeia produced with an Alpha 500 camera at the focus of a 300 mm diameter telescope at $F/D=6$; exposure : 1 minute. CCD image Cyril Cavadore, Patrick Martinez and Henri Pinna.

The beginning of the night is reserved for flat field acquisitions. If we foresee the use of filters, a series of flat fields should be carried out with these filters. For sky image acquisition, we must produce at least two exposures per observed field, while shifting the image by a few pixels in order to be able to identify, without ambiguity, later, the presence of artifacts (defective pixels, cosmic rays, etc.) or the discovery of a new object.

When observing deep sky objects in the presence of the moon, it is preferable to filter the images with a red filter which cuts the blue and green light, colors which are brighter in the sky when lit by the moon. Hence, we reduce the sky background's photon noise, therefore increasing the detectivity (at the cost of sensitivity).

6.3.3 *Multiple exposures and tracking comets*

Because of the CCD's linear response and its digital image processing facilities, it is possible to replace a unique image (exposed for 10 minutes, for example) by a series of images of the same object totaling an identical integration time (10 images exposed 1 minute each, for example). During processing, we add all of a series' images; the sum image (of 10 elementary images of 1 minute), there-fore, represents as much information as the single image (exposed for 10 minutes).

There are a few advantages to producing several images instead of one. In case of an incident (accidental guiding error, passing of an airplane or satellite in front of the object; . . .), only a single image would be affected (and rejected), and we could be content with the sum of the remaining images, while in the case of a single image, everything would have to be started over again. Otherwise, it is easy to distinguish an accidental artifact, such as a cosmic ray (which would only be present on a single image of a series) from an object in the field (which would be present on all of the images); it would be equally easy to eliminate this artifact by replacing the concerned pixels by the same pixels of another image. Finally, some observers are more comfortable with occasional guiding than staying glued to the eyepiece during the total time of the single image.

There are also some problems in producing a series of images. Each image brings with it its own read-out noise; hence, the sum of a series of images has a weaker signal to noise ratio, therefore, a worse detectivity than a single image.

There is, therefore, a compromise in terms of the exposure conditions and the camera's quality. With a camera which has very low read-out noise, we could divide a long exposure into several shorter ones: the criterion to verify, on each elementary image, is that the read-out noise stays negligible compared to the thermal noise and the sky background's photon noise. But, for a camera which is not very good for read-out noise, it would probably be worthwhile to lean toward a single exposure.

A CCD camera's manufacturer claims multiple exposures to be an advan-tage of their product: the supplied acquisition software recenters the differ-ent images of a multiple exposure before adding them. According to the manufacturer, the functioning mode's goal is to avoid guiding: each ele-mentary image is just sufficiently exposed so that the telescope's tracking imperfections stay tolerable; the sum of the images represents the total desired integration time, and the software eliminates, by percentage, the tracking imperfection(s) that would be visible from one image to another. Unfortunately, such an operating mode implies an integration time of a few seconds per image, and hence a considerable level of read-out noise

FIGURE 6.17 This image of the Swift–Tuttle comet results from the compositing of three 1 minute exposure images. The images were recentered according to the comet's nucleus.

compared to the original. The sum image, therefore, will not be as good quality as a single image, but this is not always explained to the customer . . .

But the multiple exposure can have a big advantage for the tracking of objects moving in the sky, such as comets or asteroids. In photography, comet tracking consists in removing the guide star in relation to the cross-hairs (or moving the eyepiece cross-hairs in relation to the camera) in the opposite direction to the comet's displacement and at the same speed; this operation is not simple to produce. In general, the comet's movement is only detectable with integration times above one to a few minutes. With a CCD, it is therefore advantageous to make exposures of a duration just below this critical time, then add them while recentering the comet rather than the field's stars.

6.3.4 *Image processing and analysis*

The images' pretreatment phase is extremely important: we seek to eliminate as best we can all possible artifacts: thermal changes (with the dark), non-uniform luminous response (with the flat field) and cosmetic faults. These corrections must be done all the while trying not to add noise to the images. For this, we use the dark optimization technique and the production of median flat fields.

Concerning image processing, there are no general methods as there are for planets or double stars, since deep sky objects are very different from one another. We note, however, that the goal is to obtain the least noisy and best resolved image possible, with a very uniform sky background.

FIGURE 6.18 A composite image of the central section of Messier 42 produced with an Alpha 500 camera at the focus of a 300 mm diameter telescope at $F/D=6$. Here, each exposure was limited to 5 seconds, not to correct the tracking faults, but to avoid the appearance of blooming trails on the trapezium's stars. Twelve elementary exposures were added to increase the signal to noise ratio. CCD image: C. Cavadore, P. Martinez and H. Pinna.

Besides the aesthetic aspect, CCD images of deep sky objects lend themselves well to morphological studies and large sampling statistical studies. The possible study subjects are in the dozens or hundreds; to each one corresponds a specific image treatment and analysis.

It is impossible to draw up an exhaustive list of all the possible subjects. We will content ourselves by citing a few examples:

- Study of matter jets in the coma of bright comets. These jets are weakly contrasted and almost impossible to photograph. They are accessible to CCD thanks to its large dynamic range and to the digital image processing possibilities (transform into polar coordinates to make a spherical symmetrical image of the comet, then separating this image from the real image in order to reveal the radial details).
- Polarimetric studies and comets' photometric trackings.
- Statistical study of star distributions in globular clusters.
- Statistical study of the structures of spiral galaxy arms.
- Study of elliptical galaxy isophotes. We could search, from an isophote outline, to reveal difference departures from pure ellipticity, for example, in the shape of a 'lemon' or in a 'boxy' shape.
- Research of weak matter envelopes around nebulae and galaxies; research of matter bridges between galaxies in interaction.
- Identification, in nearby galaxies, of planetary nebulae (by dividing an image centered on 500.7 nm by another in the continuum) or globular clusters (by subtracting a stellar model).

6.4 Photometry

6.4.1 *Areas of application*

Photometry consists of determining the luminous intensity of point-like objects (stars) and the brightness of extended objects (comets, galaxies, etc.). In astronomy, we use specific photometric scales: magnitude for star-like objects and surface magnitude for extended objects. The CCD image is composed of pixels in which the number of ADUs is directly proportional to the number of photons received, hence, also the luminous intensity. Since luminous intensity is linked to the magnitude by the Pogson relation, we see that it is mathematically possible to transform the ADU's scale into a magnitude scale.

Before the emergence of CCD cameras, astronomers could only carry out precise magnitude measurement with photomultiplier tubes. These tubes are composed of a photocathode which generates electrons when it receives photons. These electrons are then multiplied by a series of accelerating plates, which generate a readable signal. We can note two difficulties with the photomultiplier: the photocathode acts like a single pixel, and hence we can only measure one object at a time; and the tube can destroy itself if it is subjected to strong lighting. We describe, in table 6.1, the main advantages of CCD detectors compared with photomultiplier tubes.

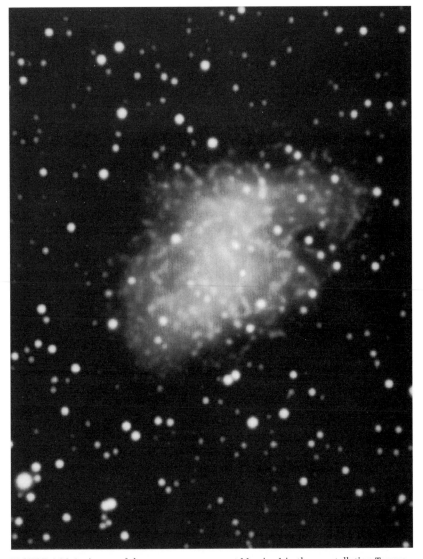

FIGURE 6.19 An image of the supernova remnant Messier 1 in the constellation Taurus, produced in white light with an Alpha 500 camera at the focus of a 300 mm diameter telescope at $F/D=6$; exposure: 5 minutes. CCD image: C. Cavadore, P. Martinez and H. Pinna.

Table 6.1

Photomultiplier	CCD
One cannot carry out a simultaneous measurement of the object and the adjacent sky background	One can carry out a simultaneous measurement of the object and the sky background
The differences in response of the detector's surface are difficult to calibrate	The differences in response of the detector's surface can be calibrated with the flat field
It is difficult to remove the interference signal contribution (cosmic rays)	Interference signals (cosmic rays) are easily correctable
No spatial information	All spatial information is available. We can easily isolate one source among many
Only one object is measurable at a time	All of the field's objects are measurable at the same time in the same atmospheric conditions
The object's centering in the diaphragm is often subjective	Centering is objective
One cannot benefit by knowing, in advance, the measured object's structure	One can benefit by knowing, in advance, the measured object's structure in order to carry out an adjustment by a mathematical model
A measurement made with a small diaphragm can be in error if the turbulence is too strong	The object continuously illuminates the detector even with strong turbulence

From a precision point of view, absolute photometric measurements carried out with a well calibrated CCD detector can be better than 0.05th magnitude.

We can distinguish two main photometric observation categories: classical photometry (stellar or surface) which consists of measuring the brightness of fixed objects reputed to have negligible luminosity variations to the scale of only a few minutes (star clusters, galaxies, etc.) and rapid photometry which consists of registering the brightness of objects subject to sudden luminosity variations and/or movements (occultations by asteroids, mutual phenomena of Jupiter's satellites, etc.).

Table 6.2

Filter	SCHOTT combination #1
B	BG39 (1 mm)+BG25 (1 mm)+GG385 (2 mm)
V	BG18 (1 mm)+GG495 (3 mm)
R	OG570 (2 mm)+KG3 (2 mm)
I	RG9 (2 mm)+RG780 (2 mm)
neutral	NG12 (4 mm)

Table 6.3

Filter	SCHOTT combination #2
B	BG12 (1 mm)+BG18 (1 mm)+GG385 (2 mm)
V	BG18 (1 mm)+GG495 (2 mm)+WG305 (1 mm)
R	OG570 (2 mm)+KG3 (2 mm)
I	RG9 (3 mm)+WG305 (1 mm)

Thanks to its high detectivity and its linear response, the CCD is capable of giving precise photometric measurements comparable to those of a photomultiplier. But it has the advantage of providing an image that we can analyze rather than a single point measurement. Thanks to its characteristics and its flexible use, the CCD is a high-performance tool in all areas of photometry: stellar photometry, surface photometry, simultaneous measurement of several objects, rapid photometry.

6.4.2 *Stellar photometry*

The use of filters CCD images destined for photometric analysis must be produced with a colored filter. We judicially choose a filter in such a way so as to be situated as close as possible to the standard photometric bands. The CCD detector allows the study of the BVR and I bands. Here are the combinations for the Schott name brand filters, calibrating the CCD to near the BVRI system. In tables 6.2–6.4, we provide three combinations from different publications (#1: French edition of this book. #2: proposed by the French AUDE association. #3: *CCD Astronomy 1995*, vol. 2, no. 4, 20–3):

The differences between these three combinations is very small. In any case,

Table 6.4

Filter	SCHOTT combination #3
U	UG1(1mm)+S8612(2mm)+WG295(1mm)
B	BG4(1mm)+BG39(2mm)+GG385(1mm)
V	BG39(2mm)+GG495(2mm)
R	OG570(2mm)+KG3(2mm)
I	RG9(2mm)+WG295(2mm)

the photometric calibration of these images (see section 4.2.3) allows us to obtain the same magnitudes regardless of the chosen combination.

It is imperative to produce flat fields for each filter. The sampling is chosen in such a way that the star image is spread over, at least, ten pixels. Typically, for a medium site, which has a turbulence of about 5 arcseconds, we sample at about 1 to 3 arcseconds per pixel.

During the image acquisition, we note the exact time of the exposure and the meteorological conditions in order to determine the air mass for each image. We must not forget to execute, at different altitudes, photometric calibration exposure fields.

The images' pretreatment must be done with extreme precision. Accordingly, we must assure the calibration field images are corrected with the same flat field and the same multiplicative constant as the images to be measured. We do not carry out a logarithmic treatment on the images to analyze. The analysis itself, consists of determining the stellar flux F, that is, the number of ADUs occupied by the star to be measured. Therefore, the flux F is expressed in ADUs, but, in practice, we often omit specifying the unit. Hence, a 100 ADU flux is often written as simply $F=100$. Since the star is always spread over several pixels, we have seen that we can use a window simulating a photometer aperture (section 4.3.2) or else execute a profile fit (section 5.7.1).

In the case aperture photometery, we measure the flux F_1 on the part of the image which contains the star then move the window to a nearby zone without a star, to measure the F_2 sky background. The star's flux F is equal to the difference of the two measurements $(F=F_1-F_2)$. We ensure that the photometer's window size is sufficiently large not to truncate the flux contained in the foot of the stellar profile (see figure 5.19). To avoid this, firstly, we display the image with close high and low thresholds, on either side of the sky background's value. Of course, we must also ensure that not a single pixel from the stellar image is saturated; it is even prudent to avoid the upper 20% of the scale to prevent an eventual non-linearity of the detector or the reading electronics.

In the case of stellar centering and measurement, it suffices to indicate the approximate position of the star to be measured; the computer itself calculates F. There are even routines which detect the position of all of the field's stars and therefore allow for automatic photometric measurement.

The magnitude constant A star's flux F is linked to a number m called the magnitude and related to F by the Pogson relation:

$$m = C - 2.5\log(F).$$

In this formula, the C coefficient is used to adjust the magnitude scale of the CCD to the standard magnitude scale. The problem of photometric calibration consists of finding the value of the constant which is called 'zero point' or 'magnitude constant'.

We have already seen how to measure a star's luminous flux F on a CCD image. The magnitude constant is simply calculated if we know the magnitude of a star from the field in which we have measured F. For example, we measure, after the sky background subtraction, a star of magnitude 0 has a flux $F = 10\,000$. We deduce that the magnitude constant is worth $C = 10.00$ ($C = 2.5\log 10\,000$). On the same image, if another star has a flux measured at $F = 300$, this signifies that it has a magnitude of 3.81 ($m = 10.00 - 2.5\log 300$). An image's calibration, therefore, only needs to know the magnitude of a single star from the field. In practice, we prefer to average the magnitude constant's value over several calibration stars, thus the interest in obtaining a calibration field.

On the first image, with a t_1 integration time, we have seen that a 0 magnitude star generates $F = 10\,000$ ADUs, and thus $C = 10.00$. If we integrated the $t_2 = 2t_1$ time image, therefore, the 0 magnitude star flux would have been 20 000 since the CCD detector is linear with time. We deduce that $C = 10.752$. Hence, we see that the magnitude constant depends on the exposure time. To know the magnitude constant value for integration time t_2, when we know C for t_1, we use the following relation:

$$C(t_2) = C(t_1) - 2.5\log(t_1/t_2).$$

Let us take the preceding example again. Suppose that, after taking this image, we make another exposure $t_3 = 3t_1$, on a neighboring field which contains no calibrated stars. We can, therefore, consider that the magnitude constant value will be the same for the new image, except for the integration time ratio connection. In these conditions, we find $C(t_3) = 11.19$. If a star's flux measurement is equal to 300, hence, it is of 5th magnitude ($m = 11.19 - 2.5\log 300$).

We have just seen that we can measure the magnitude of a field of stars which does not contain a star of known magnitude and which has a different exposure time from the calibration star image. Nevertheless, we must ensure that both observed fields were obtained under the same atmospheric stability conditions (no clouds) and the same air mass (same height above the horizon). This last problem will be taken into consideration below. On the other hand, there do not seem to be significant variations in the magnitude constant's value when there is a small variation of the temperature of the CCD $\pm 10°$.

Once we examine the Pogson formula a little closer, we see that if $F=1$ while $m=C$. This means the magnitude constant represents a star's magnitude which occupies a single pixel, in which the flux is 1 ADU above the sky background. Let us say that the magnitude constant is closely related to the limiting magnitude of the image. Nevertheless, this assertion must be moderated by the fact that the sky background is noisy and the star's image is extended over several pixels (at least 9 pixels, in general). If we indicate the standard deviation of the sky noise level as σ, we can, therefore, give an upper limit to the limiting magnitude:

$$m_{lim} C - 2.5\log(3\sigma + 10).$$

Concerning the photometrically measurable star's detection limit, we usually consider that the F flux must be above 100 ADUs. On average, we can consider that this condition is observed for stars brighter than magnitude $C-5$.

Absolute magnitude calibration Even by optimizing the choice of filters, the spectral response of a CCD+filters system is never exactly identical to the response of standard BVRI systems. Hence, the measures taken with a CCD+R filter provide a measurement located in the standard R band but, in general, which spills over a little on the I band. Hence, it is necessary to distinguish between the measured instrumental magnitude, denoted r, and the standard magnitude R. From what has just been stated:

$$r = C - 2.5\log(F) \text{ and } r = R + \alpha(R - I).$$

The α coefficient is called the color correction coefficient and $(R-I)$ the star's color index. There are also similar relations concerning the other bands (UBV and I). In practice, we determine α and C at the same time. Indeed, α and C, respectively, are equivalent to the slope and the ordinate intercept of the following equation:

$$y = C - \alpha x,$$

where $x = (R - I)$ and $y = R + 2.5\log(F)$

For N stars ($N \geq 2$) for which we know for each one of them R, $(R-I)$ and for which we have measured F, we obtain N couples (x_i, y_i) for $i=1$ to N. The α and C coefficients are obtained with the help of the following equations:

$$\alpha = \frac{\left(n\sum_{i=1}^{N} x_i y_i - \sum_{i=1}^{N} x_i \sum_{i=1}^{N} y_i \right)}{\left(\sum_{i=1}^{N} x_i^2 \right) - \left(\sum_{i=1}^{N} x_i \right)^2}; \quad C = \frac{\left(\sum_{i=1}^{N} y_i \sum_{i=1}^{N} x_i^2 \right) - \left(\sum_{i=1}^{N} x_i \sum_{i=1}^{N} x_i y_i \right)}{\left(n\sum_{i=1}^{N} x_i^2 \right) - \left(\sum_{i=1}^{N} x_i \right)^2}.$$

In order for these equations to be efficient, we need to use calibration stars whose $(R-I)$ colors are very different.

When we want to keep track of the magnitude constant's variation in relation to the zenith distance at the moment of observation, a third variable must be calculated, the extinction coefficient K defined by the following relation:

$$y = \alpha x_1 + K x_2 + C,$$

where $x_1 = (R-I)$, $x_2 = 1/\cos(z)$ and $y = R + 2.5\log(F)$.

For N stars ($N \geq 3$), for which we know for each one R, $(R-I)$ z and for which we have measured F, we obtain N triplets (x_{1i}, x_{2i}, y_i) for $i=1$ to N. The α, C, and K coefficients are obtained by solving the simultaneous equations:

$$\begin{aligned}
y_1 &= \alpha x_{1_1} + K x_{2_1} + C, \\
y_2 &= \alpha x_{1_2} + K x_{2_2} + C, \\
y_3 &= \alpha x_{1_3} + K x_{2_3} + C, \\
&\vdots \quad \vdots \quad \vdots \quad \vdots \\
y_n &= \alpha x_{1_N} + K x_{2_N} + C.
\end{aligned}$$

The magnitudes obtained by this last method are called 'magnitude outside atmosphere' since they are corrected for atmospheric extinction. Note the found α, C and K coefficients are only valid for a single filter. We must, therefore, calculate them for each filter and each night. Although it is already quite sophisticated, this method is not rigorous but it allows a measurement to a few hundredths of magnitude.

Examples from three stellar fields are presented in tables 6.5–6.7. The stars were calibrated using the Cousin BVRI photometric system. These fields (see figures 6.20–6.22) are located in M67, M92 and NGC 7790 clusters and are evenly spread in the sky so that we can observe at least one any time of the year.

Table 6.5

	B	V	R	I	(B−V)	(V−R)	(R−I)
	M67 ASD(1950)=8h 48min 37s DEC(1950)=+11° 57′ 33″						
no.							
81	9,929	10,027	10,059	10.095	−0,098	−0,032	−0,036
108	11,052	9,701	8,986	8,350	1,351	0,715	0,636
111	13,312	12,730	12,402	12,076	0,582	0,328	0,326
117	13,430	12,630	12,163	11,729	0,800	0,467	0,434
124	12,584	12,118	11,838	11,558	0,466	0,280	0,280
127	13,322	12,769	12,439	12,118	0,553	0,330	0,321
130	13,318	12,869	12,580	12,289	0,449	0,289	0,291
132	13,701	13,091	12,741	12,386	0,610	0,350	0,355
134	12,825	12,256	11,919	11,587	0,569	0,337	0,332
135	12,487	11,436	10,880	10,383	1,051	0,556	0,497
149	13,155	12,550	12,208	11,877	0,605	0,342	0,331
170		9,663	8,961	8,336		0,702	0,625
19	13,759	13,194	12,868	12,542	0,565	0,326	0,326
111	13,579	13,014	12,663	12,331	0,565	0,351	0,332
151		14,200	13,863	13,581		0,337	0,282
1198	13.729	13,152	12,819	12,524	0,577	0,333	0,295
1199	13,753	13,152	12,819	12,491	0,601	0,333	0,328
1228		12,402	11,978	11,587		0,424	0,391
1242		10,884	10,616	10,351		0,268	0,265

Table 6.6

	B	V	R	I	(B−V)	(V−R)	(R−I)
	M92 ASD(1950)=17h 15min 32s DEC(1950)=+43° 06′ 00″						
no.							
8	16,245	16,341	16,381	16,429	−0,096	−0,040	−0,048
9	16,606	16,073	15,758	15,471	0,533	0,315	0,287
10	15,398	14,628	14,159	13,662	0,770	0,469	0,497
25	16,481	15,911	15,571	15,228	0,570	0,340	0,343
26	16,352	16,416	16,420	16,419	−0,064	−0,004	0,001
100	18,157	16,999	16,292	15,560	1,158	0,707	0,732
A	14,826	14,023	13,549	13,073	0,803	0,474	0,476
AA	18,131	17,506	17,166	16,745	0,625	0,340	0,421
B	16,643	16,976	16,597	16,107	−0,333	0,379	0,490
C	18,087	17,544	17,195	16,701	0,543	0,349	0,494

Table 6.7

no.	B	V	R	I	(B−V)	(V−R)	(R−I)
9	17,072	15,414	14,440	13,487	1,658	0,974	0,953
10	16,745	16,027	15,617	15,156	0,718	0,410	0,461
16	15,947	15,330	14,971	14,559	0,617	0,359	0,412
17	16,570	16,017	15,694	15,312	0,553	0,323	0,382
21	16,688	15,977	15,537	15,029	0,711	0,440	0,508
25	16,125	15,444	15,008	14,540	0,681	0,436	0,468
100	14,834	14,391	14,139	13,824	0,443	0,252	0,315
K	14,712	13,212	12,386	11,607	1,500	0,826	0,779
S	14,796	14,252	13,910	13,507	0,544	0,342	0,403

NGC7790 ASD(1950)=23h 56min 08s DEC(1950)=+60° 54′

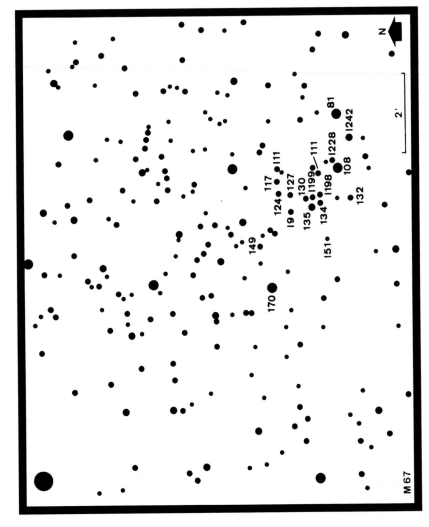

FIGURE 6.20 (refer to table 6.5)

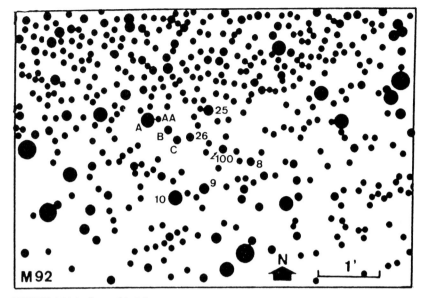

FIGURE 6.21 (*refer to table 6.6*)

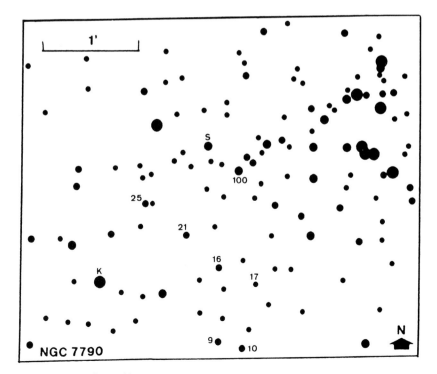

FIGURE 6.22 (*refer to table 6.7*)

The use of AAVSO and AFOEV maps Most variable star observers measure a variable star's magnitude by comparing the brightness of the star to be measured with the brightness of the calibration star present in the eyepiece's field. Variable star observation associations (AAVSO and AFOEV) provide field maps containing the calibration stars. The magnitude constant, therefore, can be determined on the variable star field itself and we naturally imagine skipping the absolute calibration. Unfortunately, the reference stars are only calibrated in V magnitude and we do not know their color index $(V-R)$. But the measurement of red stars, of the Mira type, for example, is very sensitive to this problem.

While waiting for the associations of variable stars observers to provide calibrated star maps in several colors, it would be prudent to execute an absolute calibration anyhow. The following example (see figure 6.23) concerns the V Ori and NSV 1824 variable star fields, imaged with the help of a TH7852 CCD camera equipped with a V filter and focused at $F/D=3.5$ of a 60 cm telescope at the Pic du Midi Observatory. The image was captured November 16 1992 at 0h40UT (6 minute exposure).

Table 6.8 shows values of the magnitude constant calculated with the help of four of the field's calibration stars (r,q,s,hl).

The 'star+background' and 'background' columns contain the number of measured integrated ADUs in the zones as defined for NSV1824 in figure 6.23(a). The 'star' column is the result of the difference in values of the first two columns. The magnitude constant is calculated by the formula: $C=V+2.5\log(\text{star})$. The average of the four values gives $C=21.91$ (we can estimate an inaccuracy of the order of ±0.5 magnitude).

We then use the average value of the magnitude constant to calculate the magnitude of two variable stars, with the formula: $V=C-2.5\log(\text{star})$.

Table 6.8

Name	V	Star+Background	Background	Star	C	V
r	14.9	22,576	22,163	413	21.44	
q	14.5	22,941	21,148	1,793	22.63	
s	15.1	22,499	22,176	323	21.36	
hl	12.3	48,340	39,150	9,190	22.21	
V Ori		50,444	39,300	11,144	21.91	11.8
NSV 1824		72,166	39,072	33,094	21.91	10.6

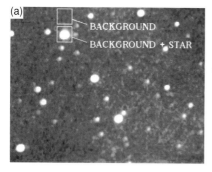

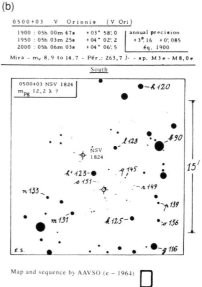

FIGURE 6.23 (a) The image of the V Ori and NSV1824 star fields with a green filter. We have indicated the intensity measuring zones of NSV1824 and the neighboring sky background. (b) A field map from the AAVSO allows certain calibration stars to be located. Image: Alain Klotz and René Roy, Association T60.

6.4.3 *Photometry of extended objects*

In the case of extended objects, such as galaxies or comets, we can distinguish two types of photometry: either we wish to know the integrated magnitude's value on the whole object (integral photometry) or else we wish to know the spatial variations of the luminescence, that is, to determine the magnitude per surface unit (surface photometry).

Integral photometry – comets We have the ability to measure a comet's global brightness in order to determine its activity parameters and its absolute magnitude. In order to do this, we must carry out integral photometry of the object.

Integral photometry stays the same as when measuring a star's magnitude: it suffices to count the total number of ADUs occupied by the extended object. In practice, we measure the total intensity of the object with a large enough aperture and we subtract the intensity measurement of all the stars which are superimposed over the object and present in the window. Finally, we must not forget to remove the sky background value. We calculate, therefore, the integral magnitude after having determined, with a calibration star, the field's magnitude constant.

There are two methods to remove the contribution of the field's stars which are superimposed on the comet image we wish to measure. The first consists of using software capable of identifying stellar images on a non-uniform background and extracting them. The second consists of taking an image of the field a few days later, once the comet has moved, and then subtracting the two images.

Before the emergence of CCD cameras, amateur astronomers determined a comet's magnitude by comparing its apparent luminosity to that of calibrated stars, whose images were defocused until they formed spots comparable in size to the comet. These methods were not very precise and needed much experience.

The determination of integrated magnitude with the help of CCD detectors is much more precise than the method of visual inspection of out-of-focus images. The photometric following of a comet's brightness could be successfully carried out with the use of CCD cameras in order to determine, for example, the degree of activity and nucleus rotation periods.

Surface photometry – galaxies The photometric study of galaxies is targeted toward the determination of spatial variation of their luminosity in order to study, in an indirect way, the distribution of stars. In surface photometry, the magnitude constant, as defined in section 6.4.2, for stars, is defined in terms of magnitude per pixel. In the case of extended objects, we usually express them in magnitude per arcsecond squared. If the CCD detector's pixels are square and the sampling's value is 'e' arcseconds per pixel, then a pixel subtends a solid angle of $A = e^2$ arcseconds squared (denoted $''^2$). For an A''^2 per pixel sampling, the surface magnitude constant C_s is deduced from C by the following relation:

$$C_s = C + 2.5 \log(A''^2).$$

On the image of a galaxy or comet, we can define a set of isophotes, each one corresponding to the limit of a given surface magnitude zone. For example, the apparent radius of a galaxy, according to De Vaucouleurs, is determined by the equivalent isophote radius of a $25''^2$ magnitude in the B photometric band. If we define an isophote as an ellipse, the equivalent radius is defined as the geometric average of the minor and major axes.

It is interesting to trace the decreasing law of a galaxy's surface magnitude according to the *Re* isophote's equivalent radius. For an elliptical galaxy, we have, in general, a linear law if we trace the surface magnitude in relation to $Re^{1/4}$.

(a)   (b)

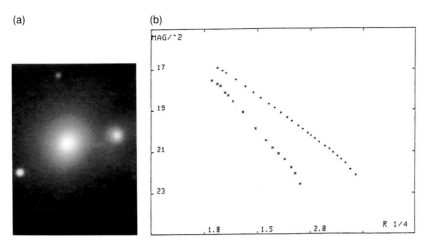

FIGURE 6.24 (a) An image of the binary galaxy ARP 167 displayed normally. (b) The surface magnitude's decreasing brightness is shown as a function of the distance from the galaxies' centers (scaled as $r^{1/4}$). Notice that the small galaxy's decrease ($\times$) is steeper than the larger one's ($+$); this is characteristic of a compact type galaxy. CCD image: Jean-Pierre Dambrine, Philippe Prugniel and Claire Scmitt-Darchy. 1 meter telescope at Pic du Midi.

6.4.4 *Rapid photometry*

In rapid photometry, the nature of the observed phenomenon sets a maximum value on the exposure time. In general, it is important to make acquisitions at a higher rate and it becomes necessary to window the image on the studied object. Also, the large number of images to save is such that the computer's hard disk can quickly saturate. It is therefore necessary to reduce the data during the acquisition, that is, to extract the total number of ADUs occupied by the object and save that value rather than the image. At the end of the observation session, the entire phenomenon can be reread on a file which has two numbers for each acquisition: the first gives the number ADUs of the object and the second the acquisition time.

The problem of tracking in time is that the object's brightness can vary for other reasons than purely astrophysical: atmospheric extinction variation, passing of clouds, etc. To quell these problems, we do two measurements on the same field, one on the variable object and the other on a supposedly stable reference star. Thus, we are doing differential photometry. It is the difference of two integrated intensities which constitute a useful measurement.

The most unfavorable case concerns the measurement of mobile and variable objects. This is not rare: mutual Jupiter satellite phenomena, tracking of an asteroid during the entire night in order to reveal its rotation period, etc. The acquisition

FIGURE 6.25 This image acquisition software screen, from the Alpha 500, shows the 'mobile double window' mode which allows the tracking of the differential brightness variations of objects moving in the field. Documentation: Mathieu Sénégas.

software, therefore, must be able to correct, at each acquisition, the position of both measurement windows, in relation to the movement of the photocenters.

6.5 Astrometry

6.5.1 *Areas of application*

The most frequent astrometry work consists of measuring the position of a comet or asteroid in relation to reference stars. Few amateur astronomers devote themselves to these measurement which are actually very useful for updating the orbital parameters of these bodies and very much in demand by professional astronomers.

6.5.2 *Image acquisition and analysis*

Traditionally, astrometric measurements are produced from photographs. For each measurement, a photograph containing the desired body and several nearby stars, which have a known position, must be produced. The position of each of these objects is then measured on the negative with a precision better than one hundredth of a millimeter. A computer program deducts the asteroid or comet's coordinates from the coordinates of the stars in the field. Two comparison stars are necessary to know the negative's orientation and scale (linked to the telescope's focal length). Indeed, we always use a larger number of stars to keep track of the optical distortion and average the measurements in which we are always looking to increase precision. This measurement phase, delicate and tedious, requiring precision material, generally distances amateurs from astrometry.

A CCD image is infinitely easier to measure than a photographic negative, since each pixel corresponds to a precise and known geometric position. Evolved astrometric functions, contained in good software, even allow a star's position to be defined with a precision superior to a pixel's, thanks to algorithms which calculate a stellar image's barycenter. The measurement operation, so delicate in the case of a photographic image, is produced in a few seconds of computer calculation on a CCD image.

The CCD, therefore, would be the miracle astrometry tool if it did not have such a limited surface area. Indeed, the precision necessary for astrometry requires that we choose a scale of, at most, 1 arcsecond per pixel (or 4 meter focal length for a Thomson 7895, which has pixels 19 μm per side). At this scale, even a 512×512 pixel CCD only covers a 8′×8′ square of the sky. Such a restricted portion of the sky, centered on the object to measure, rarely contains many bright comparison stars; but astrometrists generally use 5 to 6 per measurement. This explains the hesitance among some to abandon photography for the CCD.

In conclusion, it is necessary to obtain images with large arrays while keeping a sampling in the order of an arc second per pixel. Concerning the (x,y) position measurement of the stars in the field, we use an automatic search and analysis tool integrated in good image treatment software.

6.5.3 *Alleviating the field problem*

It is possible to avoid the handicap linked to a CCD's small size. Firstly, a CCD image allows access to fainter stars than a photographic image does. There are, therefore, a considerable number of stars on a CCD image even if its field is restrained. The problem, until recently, was that only the bright stars had positions published in catalogues (12th magnitude limit, for example, for the SAO catalogue). But the recent publication of the 'Guide Star Catalogue' on CD-ROM allows access to a much greater number of stars; the 'Guide Star Catalogue' is the guide star index used by the Hubble Space Telescope.

Also, we can foresee a procedure which allows CCD astrometry in zones where few reference stars are available. One of the advantages of the CCD is that, unlike photographic film, it benefits from a rigid support, whose geometric dimensions are constant from one image to another. It is, therefore, possible to standardize the optic (exact focal and distortion coefficients' values) during a first image of a sky zone rich in reference stars (calibrated star cluster). The CCD's orientation can then be found by letting a star drift across the field following the east–west motion of the sky (this orientation can vary slightly from one sky region to the other if the polar axis of the equatorial mounting is not properly aligned). When a single reference star is in the same field as the object to be measured, a measurement is possible. The number of asteroids accessible to a CCD and necessitating astrometric measurements is so large that there are inevitably several dozen every night which pass in proximity to a reference star; the only drawback is organizing an observation program in relation to the sky position of different asteroids and comets.

Another advantage of the CCD over photography is its greater detectivity, which renders comet and asteroid astrometric measurements possible, which would be difficult to perform with photography.

The CCD array structure and the direct computer image treatment possibilities make it a much easier tool to use in astrometry than photographic film. Its handicap comes from its small size, which means that a restricted number of reference stars are accessible in a same field. It seems, meanwhile, that this objection can be removed thanks to the fainter magnitudes reached by the CCD and the publication of new catalogues rich with faint stars.

6.6 Spectrography

6.6.1 *Areas of application*

Spectrography consists of analyzing the physical and chemical conditions of celestial bodies. For this, we use a spectrograph placed between the telescope's focus and the CCD detector. The spectrograph's goal is to separate the different 'colors' of light, that is, obtain the spectrum.

We will not describe, in detail, the functioning of different types of spectrographs. We will only study the case of a simple spectrograph constituted of a collimator, a grating, a camera objective and, eventually, a slit.

The collimator is a converging lens. Its focus is placed at the telescope's focal plane. The light beam exiting from the collimator is parallel. The grating is a thin glass plate, placed behind the collimator, engraved by micro-grooves called 'lines', whose goal is to produce diffraction and interference to cause the spectral decomposition of light. The spectrum will be in proportion to the number of dispersed 'lines' per millimeter. In general, we use gratings which have 300 or 600 lines/mm. In order to not diaphragm the light, the grating must be larger in dimension than the exiting beam diameter from the collimator.

The camera objective is made from a converging lens, for example, a photographic camera lens. Its role is to focus the light, dispersed by the grating, over the surface of the CCD detector. Hence, we form the spectral image on the CCD. To avoid diaphragming the light beam, it is necessary to use a lens whose entry diameter is at least equal to the exiting beam diameter of the grating.

When studying the spectrum of extended objects, such as the sun or nebulae, it is necessary to isolate a small slice of their image. To do this, we use a slit which is placed exactly on the plane occupied by the telescope and collimator foci.

The CCD detector collects the light intensity in relation to wavelengths. We make sure to orient the CCD's sides so that the horizontal axis is parallel to the wavelength axis (we also call it the dispersion axis).

As we disperse the light, it is necessary to have a telescope with the largest diameter possible. In photography, even with 500 mm aperture telescopes, it is very difficult to obtain good quality stellar spectrums. The CCD detector, because of its linearity, its sensitivity and its spectral extent, has given a new life to spectrography for amateur astronomers. We will not describe how to orient optics to select the proper wavelength, nor the calculation which obtains a given resolution from different optical parameters. This information can be found in specialized works (see bibliography).

A fundamental parameter is the sampling of the dispersion axis. This sampling is expressed in nanometers per pixel (it is also expressed in ångströms (Å) per pixel or 1 nm = 10 Å). We can define three sampling classes corresponding to the study of specific objects:

- Low resolution: The sampling is superior to 1 nm per pixel. Hence, we can observe faint objects showing large spectral features (novae for example).

- Medium resolution: The sampling is in the order of 0.1 nm per pixel. In this case, we can study the general structure of bright objects' spectral features (stars visible to the naked eye).

- High resolution: The sampling must be less than 0.02 nm per pixel. In this case, we can begin studying the fine structure of very bright spectral objects, that is, at the amateur level, the sun.

We use CCD arrays rather than 1-D arrays since the spectrums are never dispersed along an infinitely fine line because of turbulence and optical aberrations.

It is suggested that a binning operation be carried out in a direction perpendicular to the dispersion axis in wavelengths. For reasons cited in section 5.4.2 we often prefer carrying out a digital binning during image processing rather than during the image capture.

After the binning, the spectrum can be represented in the form of a graph in which the abscissae axis represents the wavelengths and the ordinate axis represents the intensity.

Spectrography requires long exposure times, therefore, preferably using an MPP type CCD camera. The spectrums are acquired without binning so that we can easily remove the cosmetic defects (notably the cosmic rays). The pretreatment step is important, especially the dark correction. The image treatment consists of the binning, then calibrating the wavelength axis in standard units (nanometers or ångströms) and, eventually, performing a flux measurement.

6.6.2 *Wavelength calibration*

We can only properly use a spectral correction if the dispersion axis is calibrated in wavelengths. For this, we must acquire a luminous object exhibiting very narrow spectral lines for which we very precisely know the wavelength.

The most economical solution: we calibrate the spectrums by themselves (autocalibration) from lines in which we know the associated chemical element. Unfortunately, stellar lines are always more or less shifted from their theoretical position because of their own movement and the atmospheric movement of stars (it is the values, rather, that we look for in a scientific study). Despite it all, for want of other things, it is a method which allows the identification of the main 'unknown' spectral features of a spectrum.

Intermediary solution: with the help of a halogen light, we illuminate the slit in front of which we have placed a filter. Such filters are sold by the Schott company. They are BG20 or BG36 didymium glass filters. The transmission curve must be calibrated by a precise instrument before using the filter. It is probable that their precision is not very great. Nevertheless they can, maybe, find an application in low resolution.

Expensive solution: the slit is lit by a lamp containing pure, low pressure gas emitting very fine emission lines (in general, narrower than 0.01 ångströms). The lamps currently used by professional astronomers are: the helium lamp for the visible low and medium resolutions, the argon lamp for the near infrared low and medium resolutions ($\lambda > 700$ nm) and the thorium lamp for high resolutions. Of course, it is the most expensive solution but it is, by far, the best.

The calibration in wavelength consists in finding a mathematical equation which best passes through the cloud of defined points on a graph representing, on the abscissa, the lamp wavelength lines and, on the ordinate, the pixel number corresponding to the line's centroid (in fractions of a pixel determined by approximating the line with a Gaussian). In principle, in the case of a grating, the dispersion must follow a linear law. In practice, the spectrograph's assembly of optics provokes geometric deformations that induce a law we approximate by a parabola by least squares.

In the case where calibration is done from only two argon or helium lines sufficiently separated, we assume a linear law. This is insufficient to precisely measure wavelengths on medium resolution spectrums. In low resolution, it becomes possible to measure a galaxy redshift of less than 5 Å while the sampling is of 20 Å/pixel.

6.6.3 *Chromatic calibration*

The ultimate goal of spectrography is to do spectrophotometry. The study of variable stars would be extremely detailed since we could study the continuum (like the UBVRI filters) variations as well as the temporal variations of spectral lines. Unfortunately, the study of spectrophotometry in absolute flux is never more precise than by a few percent since the exposures are long and the air mass calculation is not easy. The precise calculation of the extinction requires the observation of at least 10 calibration stars, time which could be more usefully spent observing other more 'interesting' targets! Also, well-calibrated stars are not numerous and the uncertainty of their measurement can reach 3%. Hence, spectrophotometry cannot replace classical photometry which allows superior precision.

Spectrophotometry can, however, be very useful. Despite a low precision, we can determine the color temperature of a star within 500 K without too many problems. An important rule of spectrophotometry is that the slit must be sufficiently opened so that all of the star's flux reaches the spectrograph. For this, the slit must be opened 10 arc seconds.

The spectrophotometric calibration principle is simple. We observe a standard star to determine, for each pixel of the dispersion axis, the instrumental response in $ADU/(erg/Å/cm^2)$. Then, for a spectrum measured in the same conditions, it suffices to divide by the instrumental response to deduce the object's emittance. The currently used units are: $erg/cm^2/s/Å$ or $erg/cm^2/Hz$ or even $photons/cm^2/s/Å$.

In practice, spectral image processing software is equipped with functions which allow spectral flux calibrations in a quasi-automatic way. Later, we will give the calculation's principle.

Reference star catalogues give, for about thirty wavelengths, the magnitude V as well as the magnitude per hertz, $m_{\nu\lambda}$. For $\lambda = 5445$ Å we get $m_{\nu 5445}$. At each reference wavelength, $m_{\nu\lambda}$, we calculate the absolute flux of the reference star ($F_{0\lambda}$ in erg/cm²/s/Å):

$$F_{0,\lambda} = 1108(10^{-0.4 \cdot (Vm_{\nu\lambda} - m_{\nu 5445})})/\lambda^2,$$

where the wavelength is expressed in ångströms. On the observed spectrum of the standard star, we know t_0, the exposure time in seconds and e_0, the sampling of a pixel (in Å/pixel). For each of the pixels associated with the wavelength λ, we measure the number of ADUs $f_{0,\lambda}$ and calculate the instrumental response:

$$K_\lambda = f_{0,\lambda}/(t_0 e_0 F_{0,\lambda}) \text{ in ADU/(erg/cm}^2)$$

Then, the K_λ values for all of the spectrum's pixels must be interpolated. To do this, we use the Spline functions.

For whatever object, observing with an exposure time t (in seconds) with an e sampling (in Å/pixels), each f_λ flux pixel (in ADUs) is divided by K_λ from the spectral response. The result, F_λ, represents the absolute flux of the observed object:

$$F_\lambda = f_\lambda/(te K_\lambda) \text{ in erg/cm}^2/\text{s/Å}.$$

In practice, we must remember, before beginning calculations, to correct the $f_{0,\lambda}$ and f_λ fluxes for the atmospheric extinction in order to place ourselves in 'outside atmosphere' conditions.

6.6.4 *Factors limiting performance*

Focusing is the most important factor. Indeed, poorly adjusted optics can easily lead to a 100% drop in efficiency! The adjusting of a spectrograph must be done this way and in the following order:

1 The camera objective is adjusted for infinity focus,
2 The collimator is focused on the slit,
3 The slit is placed at the prime focus of the telescope.

For point no.1, it is adequate, before installing camera objective+CCD assembly onto the spectrograph, to point at an object several kilometers away. Hence, we set the image acquisitions in automatic mode to obtain a series of short exposures while adjusting the lens's focus to obtain the cleanest image possible. We are obliged to use this method with CCD cameras since the distance indicators on the lens often do not correspond to reality.

For point no.2, the objective +CCD assembly, now adjusted, is installed onto the spectrograph. The slit is adjusted as finely as possible and we light it with a low pressure gas lamp (calibration lamp). We make automatic image acquisitions, while slowly moving the collimator, in such a way to obtain the finest possible emission lines.

For point no.3, we target a type A blue star (Vega for example), we open the slit very wide (1mm for example) and adjust the focus on the Hα line. The slit +collimator assembly must be linked by a translation plate. We slowly move the plate so as to obtain the broadest possible line.

Because of turbulence and optical aberrations, the spectrum extends itself several lines in height which must be 'picked up' in a single (hence, we add according to the columns) binning operation. Therefore, we improve the signal to noise ratio. The problem consists of knowing to what extent the binning must be done. We show that the best signal to noise ratio is obtained when we only add the intensity lines which are brighter than about two times the sky background's noise amplitude.

The slit is an element which must be cared for. Geometric imperfections can lead to flux variations of the order of 50% (or more) according to their size. The slit must always be kept clean (not touching it is still the best solution). Another solution consists of buying an aluminized glass plate on which is masked the desired slit dimension.

6.6.5 *Analysis software*

A spectral data analysis software must be able to carry out the following operations:

- Binning of the spectral image to make a profile,
- Wavelength calibration,
- Photometric calibration,
- Measurement of central wavelengths of spectral features,
- Intensity line measurements and their equivalent width,
- Addition, sum and subtraction of spectrums,
- Adjusting in wavelengths a spectrum in relation to another,
- Spatial convolution (high-pass and low-pass filters),
- Fourier deconvolution with the instrumental function.

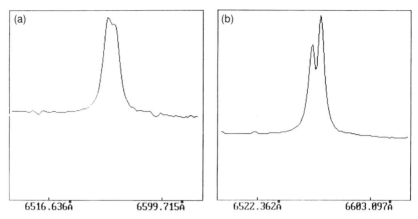

FIGURE 6.26 (a) The spectrum of the star Taurus ζ obtained on 25/2/91, shows the Hα line in emission. (b) the same star, on 4/3/93, shows an evoluton of the absorption structure. Image: D. Bardin, V. Desnoux and M. Espoto. Association T60.

6.6.6 *Example application*

In medium resolution, we can see the evolution, from year to year, of the Hα line profile of Be stars. Therefore, we can follow the progression of the different envelopes of these stars, a phenomenon still not well understood today.

In low resolution, we can attempt the detection or tracking of emission lines of non-periodic variable stars (red giants, novae, Be, etc.). It should be possible to obtain the spectrum of supernovae in galaxies of the Virgo cluster ($V=12$ at maximum) and therefore, determine the type of supernovae without ambiguity (the type Ia represents a large absorption toward 6500 Å while type II shows a large Hα emission).

From an enthusiasts point of view, it is very easy to study the spectral features of a given wavelength range in relation to the type of star, to determine the surface temperature of stars, to observe and measure the intensity of forbidden lines of planetary nebulae, identify methane on gaseous planets or on Titan, etc.

6.7 Sky surveillance

6.7.1 *The area of study*

The goal of sky surveillance is to discover a new object or any other unexpected changes: novae or supernovae explosion, new asteroid or comet, etc.

On a strategic observation plan, we can arrange the different objects to research into two categories:

- Those that are searched for in restricted areas of the sky: they are typically extra-galactic supernovae, which can only appear in the field of one galaxy. There are also periodic comets whose return is expected: we roughly know where the comet will be, on the return path toward the sun, when it will be observable from Earth.

- Those that must be searched for over the entire or on a large part of the celestial sphere: they are the new long period comets, which can appear in any direction, the new short period comets and asteroids, which are searched for in a certain band around the elliptic and the novae which frequent the Milky Way.

This distinction is fundamental for the CCD observer, in which the weak point compared to the visual observer or the photographer is the small field covered by its detector. It is clear that the CCD suffers little from this inconvenience for objects researched on a small field. But, the question must be seriously studied when we must cover a considerable part of the sky. We have, therefore, saved this classification for the present analysis.

6.7.2 *Limited zone searches*

Supernova searches A few dozen supernovae are discovered each year. Most of the time, these supernovae are discovered on photographic plates taken by the large professional Schmidt telescopes. The problem is, on the one hand, reaching a sufficiently faint magnitude to detect a supernova in a not too distant galaxy (most supernovae are discovered between the 13th and 19th magnitude) and, on the other hand, covering a large number of galaxies: if we count an average of 1 supernova per galaxy every 30 years, we must statistically explore 3000 galaxies to find 1 supernova which has exploded within the last 3 or 4 days.

The Schmidt telescopes cover large sky zones using photographic plates. In regions rich in galaxies, it is possible to survey a large number of galaxies on each image. Contrarily, a CCD camera mounted on a Newton telescope only covers one or a few galaxies at a time. In return, the CCD is much more convenient for the analysis of images. It is possible to examine the galaxy in the seconds following the end of the integration time. The visualization functions of available software easily allows the revealing of stars even if they are close to the galaxy's nucleus and centered in its luminosity. Whereas in photography, these stars would be difficult to see.

An exposure of a few seconds with a CCD camera mounted on a medium-sized amateur telescope is adequate to reach a sufficiently faint magnitude limit.

The observation program must plan neighboring galaxy sequences in order for the movement from one galaxy to another to require only a small movement on the sky. If the telescope is equipped with an efficient pointing method, it is then possible to jump from one galaxy to another in one to two minutes. While an operator does the pointing to the next galaxy, another member of the team can examine the image of the preceding galaxy on the computer screen. Some software, such as the one for the Alpha cameras, allows the side by side displaying of the image taken and the comparison image, which permits easy examination or even the superimposing in blink mode (see section 6.2.2) to more easily reveal a supernova. We can even imagine software which automatically carries out the search for stars and checks them to see if they are new by comparing them with an archive file. Even with a 'manual' supernova search, it appears possible to cover, by CCD, almost a dozen galaxies an hour, up to stellar magnitudes of 18 or 19. In these conditions, the CCD can be competitive compared to Schmidt telescopes, especially in sky regions where the galaxy density is not too high. It is even suggested that galaxy clusters be avoided in order to be complementary rather than a rival to Schmidt telescopes.

The rediscovery of periodic comets Periodic comets are only visible in the part of their orbit closest to the sun and Earth. Since the orbits of these comets are known, we can predict their return. About ten to twenty periodic comets are awaited each year. It is, therefore, interesting to locate the comet as early as possible while it is still far enough from the sun and very faint. A few astronomers in the world are specialists in this sport. The recoveries are at about 20th magnitude, which is achievable by astronomers equipped with a medium-sized telescope and a good camera.

The expected returns of comets are published by the International Comet Quarterly. If the comet was not widely observed in the past, it is probable that its position forecast is erroneous; the area around the expected position, therefore, should be investigated. In general, the orbit's geometric parameters are quite reliable, but the comet may be early or late; we must, therefore, search along the trajectory. A small ephemeral program inserted into a computer would be useful to calculate the comet's position in the sky each day if it is on the forecast orbit but a few days early or late.

It is necessary to operate under a very dark sky, with at least a 30 cm telescope and a CCD camera with a low dark current. Because of the searched-for comets' smallness and the large number of objects present at the magnitude reached, it is necessary to have a good resolution (not more than 4″ per pixel, thus requiring a long focal length. For example, an Alpha 500 camera used with a 300 mm Newton telescope open at $F/D=6$ gives a resolution of 2″ per pixel

FIGURE 6.27 An image of the galaxy Messier 74, in the constellation Pisces, produced with an Alpha 500 camera at the focus of a 300 mm telescope at $F/D=6$; exposure: 5 minutes. CCD image: Cyril Calvadore, Patrick Martinez and Henri Pinna.

Table 6.9

NGC	Number of supernovae	Magnitude of the brightest	Last discovery
224	1	V 5.9	1885
991	1	B14.0	1984
1003	1	12.9	1937
1058	2	12.7	1969
1073	1	B14.0	1962
1316	2	B12.5	1981
1433	1	B13.5	1985
1511	1	12.5	1935
LMC	1	B4.5	1987
2608	1	11.8	1920
3184	3	11.0	1931
3198	1	B11.4	1966
3913	2	B12.3	1979
3938	2	13.4	1964
4214	1	9.8	1954
4254	3	14.0	1986
4303	3	12.0	1964
4321	4	B11.6	1979
4382	1	B11.9	1960
4486	1	12.3	1919
4496	2	B11.6	1988
4536	1	B12.0	1981
4564	1	B11.8	1961
4618	1	B12.1	1985
4621	1	12.0	1939
4636	1	12.6	1939
IC4182	1	8.4	1937
5055	1	B11.9	1971
5236	5	B11.6	1983
5253	2	8.0	1972
5457	3	B11.7	1970
6384	1	B13.6	1971
6946	6	B11.6	1980

and allows the location of a 20th magnitude comet of 15″ in diameter in a 30 minute exposure.

If the comet's movement in the sky is greater than ten arc seconds per hour, it is useful to perform telescope guiding based on the calculated comet movement (or an adequate recentering on a series of images); the stars will be slightly streaked, but the comet image will stay on the same pixels, which is favorable for detection.

Once a suspected object is detected, several images must be produced at one or several hours of interval to ensure it is not an artifact and confirm that it is a comet by its movement.

If we become enthused enough to try and be the first to rediscover a comet, it is best to attempt comets which have a low elongation upon their return; we must, therefore, point near the horizon, which handicaps professional telescopes more than amateur ones. Once the comet is identified with certainty, we must make a precise astronomical measurement of its position and telegraph it without delay to the Central Bureau for Astronomical Telegrams of the International Astronomical Union.

Thanks to its heightened detectivity and its computer image analysis possibilities, the CCD brings an efficiency superior to photography when searching for objects in a restricted field (supernovae, periodic comet rediscoveries).

6.7.3 *Extended zone searches*

The more the searched-for objects are faint, the more we need a large telescope and a long exposure time. But the more powerful the telescope is, the more its field is reduced, which implies more time to cover the celestial sphere.

The problem presents itself in the following terms. The visual observer who uses a 200 mm telescope reaches 13th magnitude stars and 11th magnitude comets. The maximum field according to the telescope is in the order of 2°; we can, therefore, consider that it is capable of surveying 2000 square degrees of sky per hour of observation (on condition it does not lose time searching on an atlas all of the galaxies encountered, which supposes that the sky is known by heart); but the celestial sphere covers just over 41000 square degrees! If this observer decides to equip him- or herself with a 310 mm telescope, he or she could reach 12th magnitude comets, but the maximum observable field is in the order of 1.3° and will not cover more than 850 square degrees per hour of observation.

It is the same problem in photography. A 13 cm diameter Schmidt camera reaches a 14th comet magnitude and covers 50 square degrees per hour of work (image capturing and developing) while a 20 cm diameter Schmidt camera reaches a 15th magnitude comet and only covers 30 square degrees per hour of work.

To present the same resolution as a Technical Pan type film used in a 20 cm diameter Schmidt camera with a focal length of 30 cm already taken in the previous example, a camera such as the Alpha 500 (which uses one of the largest CCDs available to amateurs) must be used with a 60 cm focal length; it could be, for example, a 15 cm diameter Newton telescope at $F/D=4$. The field covered by each block is only 0.8 square degrees. But a 1 minute exposure is enough to identify a 15th magnitude comet. If the operator is well organized, he or she could maybe capture an image every 2 minutes (one minute for the acquisition and one minute for the pointing) and, aided by the camera software's 'blink' function, analyze the image in a minute. We see that the coverage can be 16 square degrees per hour worked, which is two times worse than with the Schmidt camera. But, the work is less annoying.

In this race against time, which consists of surveying the largest portion of sky possible, up to the faintest magnitude possible, in the shortest time, CCDs present the major drawback of covering a tiny field, but the advantage of a good detectivity and, mostly, the bonus of high-performance digital image processing.

Nothing resembles an asteroid or nova more than a star among hundreds of thousands of stars accessible to amateur telescopes; nothing resembles a comet more than a galaxy among the thousands of galaxies accessible to amateur telescopes. We must, therefore, be capable of quickly telling whether the suspected object is or is not part of the known objects catalogue; its eventual movement, in the case of an asteroid or comet will, eventually, confirm its nature.

One of the most useful software functions is the 'blink'. This function consists of superimposing the image to be explored on a reference image and alternately displaying them one after another on the screen, several times a second. Objects which are on both images appear fixed, but any new object blinks and immediately attracts the operator's attention. A 'blink microscope' is an optical apparatus which allows the superimposition, in the eyepiece, of two photographic negative images which are alternately lit; this apparatus is very useful in the search for new objects on photographs of the sky, but is difficult to make for an amateur astronomer. But good image processing software has a particularly easy 'blink' function to use with CCD images.

6.7.4 *A particular case–TDI mode*

The CCD can hope to compete with photography in the surveying of extended sky zones thanks to two of its specific possibilities: TDI mode and the automated processing of images.

TDI mode consists of letting the sky pass before the fixed telescope. At the telescope's focus, a CCD array is read line by line with a transfer speed that corresponds to the sky's image displacement on the photosensors; hence, we produce a type of electronic tracking of the image. The other characteristic of TDI mode is to provide an image 'to the kilometer': the height of the image, in declination, is equal to the angle on the sky covered by the length of a CCD line; its length, in right ascension, is equal to the passing of the sky before the telescope during the duration of the image. The integration time is equal to the time it takes the sky to pass from the first to last line of the CCD.

Placed at the focus of the 150 mm diameter telescope with a 600 mm focal distance of the preceding example, an Alpha 500 CCD camera covers $0.9°$ in declination. At the celestial equator, the sky passes at $15°$ per hour; hence, the field covered by the TDI image is 14 square degrees per hour. The integration time is 220 seconds, which allows the detection of 16th magnitude comets and 19th magnitude asteroids. It is, therefore, useful to develop software which analyzes the TDI image and automatically compares it to a reference file. Thus, we use a powerful sky surveying method, which is capable of detecting very faint objects, but at a relatively slow rate. It is placed, therefore, in a very different category from the traditional investigation field of amateur comet researchers; but one of the secrets of success in comet research is to do what others are not. Remember that because of the angular sky speed variation in relation to the declination, TDI mode is only useful near to the celestial equator.

In a search for objects over large areas the CCD suffers greatly from its small size compared to the traditional methods (visual, photographic). Particular procedures, which have yet to be explored, should give it a good efficiency in certain categories, notably for searching for very faint objects. The TDI mode and expert form identification software should help the CCD on this path.

Conclusion

We hope that these pages have convinced you of the performance and simplicity of CCD imagery. We also hope that they have shown you what a CCD camera is and given you the essential ideas to equip yourself and use that equipment well.

This book is not an end in itself. Rather, it is one of the pieces of a puzzle which has barely begun to assemble itself. It represents only an introduction to amateur CCD astronomy. The subsequent chapters of this story will be written by the observers themselves. The ADAGIO association hopes to participate in this adventure, through the organization of new workshops and symposia, through its observational work and publications. It hopes to maintain links with the readers of this book to exchange results and information and to continue guiding if necessary.

At the time of this book's publication, amateur CCD astronomy is at its beginnings. It will blossom in the years to come. It will then be time to take stock and examine what actions will enable the amateur observer community to take full advantage of this new tool.

Bibliography

GENERAL WORKS

R. Berry, *Introduction to Astronomical Image Processing*, William-Bell Inc., 1991. Book and disk containing the AIP software and giving a basic introduction to image processing.

R. Berry, *Choosing and Using a CCD Camera*, William-Bell Inc., 1991. Book and disk containing the QwikPIX software and giving a familiarization of image processing.

C. Buil, L'utilisation d'une matrice CCD, étude CNES, 1986. Article containing information on how to build the electronics of a CCD camera.

C. Buil, *CCD Astronomy – Construction and Use of an Astronomical CCD Camera*, William-Bell Inc., 1991. This book deals with the construction of CCD cameras up until their use in astronomy, while also dealing with image processing techniques.

S. B. Howell, *Astronomical CCD Observing and Reduction Techniques*, Astronomical Society of the Pacific Conference Series.

P. Léna, *Méthodes physiques de l'observation*, Inter-édition Edition du CNRS, 1986. This book covers a wide range of detector types and places the CCD among them.

P. Martinez *et al.* (collective work) *The Observer's Guide to Astronomy*, Volumes 1 and 2, Cambridge University Press, 1994. This book is an ideal complement for data analysis of CCD images. A chapter is devoted to CCD cameras.

WORKS AND PUBLICATIONS SPECIALIZING IN IMAGE PROCESSING

A Jain, *Fundamentals of Digital Image Processing*, Prentice-Hall, 1989. This book is a colossal mass of various image processing algorithms. Nevertheless, the formal presentation used requires a good knowledge of mathematics.

J. Lorre, *Digital Image Processing in Remote Sensing*, Taylor & Francis Inc., 1988, pages 245–269. Interesting algorithms and illustrations on point to point operations in color space and polarizing space are presented.

A. Marion, *Introduction aux technique de traitment d'images*, éd. Eyrolles, 1987. This book clearly reveals the problems of convolutions and the benefits of the Fourier transformation.

J. Skilling and R. Bryan, 'Maximum Entropy Image Reconstruction: general algorithm', *Monthly Notices of the Royal Astronomical Society*, 1984, Volume 211, pages 111–124. a report on the different maximum entropy methods, from the simplest to the most complicated.

R. White and R. Allen, The Restoration of *HST* Images and Spectra, Space Telescope Science Institute, 1990. This work outlines over twenty image restoration algorithms.

LISTS AND PHOTOMETRIC CALIBRATION STAR FIELDS

C. Christian *et al.*, 'Video Camera/CCD Standard Stars', *Publication of the Astronomical Society of the Pacific*, 1985, volume 97, pages 363–372. Six star clusters are calibrated in BGRI.

M. Joner and B. Taylor, 'Cousins VRI Standard Stars in the M67 dipper Asterism', *Publications of the Astronomical Society of the Pacific*, 1990, volume 102, pages 1004-1017. About twenty stars in Messier 67 are calibrated in BGRI.

A. Landolt, 'UBVRI Photometric Standard Stars in the Magnitude Range 11.5<V<16.0 Around the Celestial Equator', *The Astronomical Journal*, 1992, volume 104, page 340. a list of 526 calibrated stars distributed along the celestial equator.

ASTROMETRIC CALIBRATION STAR FIELDS

G. Gatewood *et al.*, 'A Study of Upgreen 1', *The Astrophysical Journal*, 1988, volume 332, pages 917–920. This cluster of seven stars allows the calibration of field distortion parameters.

SPECTROGRAPHY

D. Bardin and A. Klotz, 'Le spectrographe du T60', Pulsar No.678, 1990, pages 84–91. A general article on the characteristics of a spectrograph dedicated to CCD observations.

V. Desnoux and M. Espoto, 'Comète Shoemaker-Levy 1991a1', outlined during the Recontres de Chinon 1993, ATCO special No37, pages 43–49. A general outline on the observation and analysis techniques of spectra in CCDs.

Tug and White, 'Absolute energy distributions of α Lyrae and 109 Virginis from 3295 Å to 9040 Å', *Astronomy and Astrophysics*, 1977, volume 61, pages 679–684.

Hayes and Latham, 'Spectrophotometry of Vega between 6800 and 10400 Å' *The Astrophysical Journal*, 1975, volume 197, pages 587–592. Both of these articles permit spectral calibration in flux with the help of the star Vega.

FITS FORMAT DEFINITION

NASA/OSSA Office of Standards and Technology, *Definition of the Flexible Image Transport System*, 8 March 1993, ed. NASA Goddard Space Flight Center, Greenbelt MD 20771, USA. A hundred page document giving the complete, official definition of the FITS format, regularly updated.

A. Klotz, 'Lettre à AUDE (2): le format FITS', Pulsar No.690, 1992, pages 94–96. A summary explanation of the FITS format.

Index